SpringerBriefs in Business

For further volumes:
http://www.springer.com/series/8860

Elias G. Carayannis • David F.J. Campbell

Mode 3 Knowledge Production in Quadruple Helix Innovation Systems

21st-Century Democracy, Innovation, and Entrepreneurship for Development

 Springer

Elias G. Carayannis
George Washington University
Washington, DC, USA
caraye@gwu.edu

David F.J. Campbell
Alpen-Adria-University Klagenfurt
and
University of Applied Arts
Vienna, Austria
david.campbell@uni-klu.ac.at
david.campbell@uni-ak.ac.at

ISSN 2191-5482   e-ISSN 2191-5490
ISBN 978-1-4614-2061-3   e-ISBN 978-1-4614-2062-0
DOI 10.1007/978-1-4614-2062-0
Springer New York Dordrecht Heidelberg London

Library of Congress Control Number: 2011942511

Printed on acid-free paper

Springer is part of Springer Science+Business Media (www.springer.com)

Contents

**Mode 3 Knowledge Production in Quadruple
Helix Innovation Systems** 1

1   Introduction to Knowledge and Definition of Terms 1
     1.1   Open Innovation Diplomacy, Quadruple Helix
         Innovation, "Mode 3" Knowledge Production System,
         and Fractal Research, Education and Innovation Ecosystem
         (FREIE) 2
     1.2   Definition of Terms 6
     1.3   Mode 3, Quadruple Helix, Quintuple Helix, Democracy
         of Knowledge, Schumpeter's Creative Destruction, and
         the Co-evolution of Different Knowledge Modes 12
2   The Conceptual Understanding of Knowledge and Innovation 28
     2.1   Innovation Placed in Context 28
     2.2   The Relationship Between Knowledge and Innovation 30
     2.3   The "Mode 3" Knowledge Production System Multilevel
         Approach to Knowledge and Innovation: The "Multilevel
         Innovation Systems" 32
     2.4   Linear Versus (and/or) Nonlinear
         Innovation Models (Modes) 35
     2.5   Extending the "Triple Helix" to a "Quadruple Helix"
         Model of Knowledge and Innovation 37
     2.6   Coexistence and Co-evolution of Different Knowledge
         and Innovation Paradigms 39
     2.7   The "Co-opetitive" Networking of Knowledge Creation,
         Diffusion, and Use 41

3   Sectoral Systems of Innovation and Technology Dynamics   41
    3.1   Sectoral Systems of Innovation and Technology
          Dynamics in IT/ICT (Information Technology/Information
          and Communication Technology) in the Software Sector   42
    3.2   Sectoral Systems of Innovation and Technology
          Dynamics in Life Sciences, Biotechnology, and
          the Pharmaceutical Industry   44
    3.3   Sectoral Systems of Innovation and Technology
          Dynamics in the Machine Tool Sector   46
4   Conclusion   47
References   56

Mode 3 Knowledge Production in Quadruple Helix Innovation Systems

Twenty-first-Century Democracy, Innovation, and Entrepreneurship for Development

Abstract Developed and developing economies alike face increased resource scarcity and competitive rivalry. In this context, science and technology appear as an essential source of competitive and sustainable advantage at national and regional levels. However, the key determinant of their efficacy is the quality and quantity of entrepreneurship-enabled innovation that unlocks and captures the benefits of the science enterprise in the form of private, public, or hybrid goods. Linking basic and applied research with the market, via technology transfer and commercialization mechanisms, including government–university–industry partnerships and capital investments, constitutes the essential trigger mechanism and driving force of sustainable competitive advantage and prosperity. In this volume, the authors define the terms and principles of knowledge creation, diffusion, and use, and establish a theoretical framework for their study. In particular, they focus on the "Quadruple Helix" model, through which government, academia, industry, and civil society are seen as key actors promoting a democratic approach to innovation through which strategy development and decision-making are exposed to feedback from key stakeholders, resulting in socially accountable policies and practices.

Keywords Entrepreneurship • Innovation • Knowledge cluster • Knowledge management • Mode 3 • Quadruple Helix • Research and development (R&D) • Science and technology policy (S&T policy) • Triple Helix • Quintuple Helix

1 Introduction to Knowledge and Definition of Terms

New frontiers of the mind are before us, and if they are pioneered with the same vision, boldness, and drive with which we have waged this war we can create a fuller and more fruitful employment and a fuller and more fruitful life.

Franklin D. Roosevelt
November 17, 1944.

E.G. Carayannis and D.F.J. Campbell, *Mode 3 Knowledge Production in Quadruple Helix Innovation Systems*, SpringerBriefs in Business 7, DOI 10.1007/978-1-4614-2062-0_1, © Elias G. Carayannis and David F.J. Campbell 2012

1.1   Open Innovation Diplomacy,[1] Quadruple Helix Innovation,[2] "Mode 3" Knowledge Production System,[3] and Fractal Research, Education and Innovation Ecosystem (FREIE)[4]

Our conceptual point of departure here is our article release in the International Journal of Technology Management (IJTM) that was published back in 2009: "*Mode 3*" and "*Quadruple Helix*": *Toward a 21st Century Fractal Innovation Ecosystem* (Carayannis and Campbell 2009). In the following, we iterate and reiterate our earlier work and focus on analytically and discursively expanding our previous propositions. With this analytical expansion we want to reflect the discussions since. We also want to develop a more future-oriented outlook and vision, addressing the current challenges and introducing a problem-solving that is interested in sustainable solutions, emphasizing a sustainable development perspective that brings together *innovation, entrepreneurship, and democracy.*[5]

Developed and developing economies alike face increased resource scarcity and competitive rivalry. Science and technology increasingly appear as a main source of competitive and sustainable advantage for nations and regions alike. However, the key determinant of their efficacy is the quality and quantity of entrepreneurship-enabled innovation that unlocks and captures the pecuniary benefits of the science enterprise in the form of private, public, or hybrid goods. In this context, linking university basic and applied research with the market, via technology transfer and commercialization mechanisms, including government–university–industry partnerships and risk capital investments, constitutes the essential trigger mechanism and driving device for sustainable competitive advantage and prosperity. In short, university researchers properly informed, empowered, and supported are bound to emerge as the architects of a prosperity that is founded on a solid foundation of scientific and technological knowledge, experience, and expertise and not on fleeting and conjectural "financial engineering" schemes. Building on these constituent elements of technology transfer and commercialization, *Innovation Diplomacy* encompasses the concept and practice of bridging distance and other divides (cultural, socio-economic, technological, etc.) with focused and properly targeted initiatives to connect ideas and solutions with markets and investors ready to appreciate them and nurture them to their full potential.

We believe that the top universities are—perhaps more informally than not—already enacting a Mode 3 modus operandi as well as experimenting with Quadruple and even Quintuple Innovation Helix structures, mandates, policies, and practices. However, we are indeed calling for a more explicit and coherent strategy—per the

---

[1] See Carayannis, BILAT, March 2011, SAIS TRC, June 2011 and Springer JKEC, Fall 2011.
[2] See Carayannis and Campbell, IJTM, 2009.
[3] See Carayannis and Campbell, IJTM, 2009.
[4] See Carayannis, BILAT, March 2011, SAIS TRC, June 2011 and Springer JKEC, Fall 2011.
[5] See also Carayannis and Campbell 2011.

FREIE concept, explained in greater detail below—which is already embedded in emerging white papers and other policy documents as well as practice guidelines in developed and developing countries.

The emerging *gloCalizing*, globalizing, and localizing (Carayannis and von Zedwitz 2005; Carayannis and Alexander 2006) frontier of converging systems, networks, and sectors of innovation that is driven by increasingly complex, nonlinear, and dynamic processes of knowledge creation, diffusion, and use confronts us with the need to reconceptualize—if not reinvent—the ways and means by which knowledge production, utilization, and renewal take place in the context of the knowledge economy and society (*gloCal knowledge economy and society*).

Perspectives from and about different parts of the world and diverse human, socio-economic, technological, and cultural contexts are interwoven to produce an emerging new worldview on how specialized knowledge, that is embedded in a particular socio–technical context, can serve as the unit of reference for stocks and flows of a hybrid, *public/private*, *tacit/codified*, *tangible/virtual good* that represents the building block of the knowledge economy, society, and polity.

"Mode 1" of *knowledge production* refers primarily to basic university research (basic research performed by the higher education sector) that is being organized in a disciplinary structure. "Mode 2" focuses on knowledge application and a knowledge-based problem-solving that involves the following principles: "knowledge produced in the context of application"; "transdisciplinarity"; "heterogeneity and organizational diversity"; "social accountability and reflexivity"; and "quality control" (Gibbons et al. 1994; see also Nowotny et al. 2001, 2003, 2006). As a more far-reaching reconceptualization of knowledge production we postulate and introduce a new approach that we call the *"Mode 3" Knowledge Production System* (expanding and extending the "Mode 1" and "Mode 2" knowledge production systems), which is at the heart of the *FREIE*, and consisting of "Innovation Networks" and "Knowledge Clusters" (see definitions below) for knowledge creation, diffusion, and use (Carayannis and Campbell 2006a). This is *a multilayered, multimodal, multinodal, and multilateral system*, encompassing mutually complementary and reinforcing innovation networks and knowledge clusters consisting of human and intellectual capital, shaped by social capital and underpinned by financial capital.

The Mode 3 Knowledge Production System architecture focuses on and leverages higher order learning processes and dynamics that allow for both top-down government, university, and industry policies and practices and bottom-up civil society and grassroots movements initiatives and priorities to interact and engage with each other toward a more intelligent, effective, and efficient synthesis. In so doing, Mode 3 ensures a tighter and more robust coupling of vision with reality and helps reify the socio-economic and socio-political being and becoming by achieving between aspirations and limitations. For instance, a case in point of a Mode 3 Knowledge Production System is in the Swiss referendum system where immediate democracy shapes and drives government, academic, and industrial policies and practices and where a proper calibration of the issues addressed and the frequency modulation of the feedback received via the referenda allows for higher order learning to impart intelligence and enact wisdom in choices and initiatives.

Mode 3 per the comments above is the knowledge production system architecture that engages actively higher order learning (learning, learning-to-learn, as well as learning-to-learn-how-to-learn [Carayannis, doctoral thesis 1994; Carayannis 2001]) in a multilateral, multimodal, multinodal, and multilayered manner involving thus entities from government, academia, industry, and civil society as well as driving co-opetition, co-specialization, and co-evolution resource generation, allocation, and appropriation processes that result in the formation of modalities such as innovation networks and knowledge clusters. These modalities form as they represent topologically and thematically optimal resource agglomeration and leveraging schemes in the context of a Mode 3 knowledge production system architect.

The "Mode 3" Knowledge Production System is in short the nexus or hub of the emerging twenty-first century Innovation Ecosystem,[6] where *people,*[7] *culture,*[8] *and technology*[9,10] (Carayannis and Gonzalez 2003;—forming the essential "Mode 3" Knowledge Production System building block or "knowledge nugget" [Carayannis 2004]) meet and interact to catalyze creativity, trigger invention, and accelerate innovation across scientific and technological disciplines, public and private sectors (government, university, industry, and nongovernmental knowledge production, utilization, and renewal entities as well as other civil society entities, institutions, and stakeholders), and in a top-down, policy-driven as well as bottom-up, entrepreneurship-empowered fashion. One of the basic ideas of the article is *co-existence, co-evolution,* and *co-specialization* of different knowledge paradigms and different knowledge modes of knowledge production and knowledge use as well as their co-specialization as a result. We can postulate a dominance of knowledge heterogeneity at the systems (national and transnational) level. Only at the subsystem (subnational)

---

[6] Furthermore, see Milbergs (2005).

[7] See discussion on democracy in the conclusion of this article.

[8] "*Culture* is the invisible force behind the tangibles and observables in any organization, a social energy that moves people to act. Culture is to the organization what personality is to the individual—a hidden, yet unifying theme that provides meaning, direction, and mobilization." (Killman 1985).

[9] *Technology* is defined as that "which allows one to engage in a certain activity ...with consistent quality of output," the "*art of science and the science of art*" (Carayannis 2001) or "*the science of crafts*" (Braun 1997).

[10] We consider the following quote useful for elucidating the meaning and role of a "*knowledge nugget*" as a building block of the "Mode 3 Innovation Ecosystem": "People, culture, and technology serve as the institutional, market, and socio-economic 'glue' that binds, catalyzes, and accelerates interactions and manifestations between creativity and innovation as shown in Fig., along with public-private partnerships, international Research & Development (R&D) consortia, technical / business / legal standards such as intellectual property rights as well as human nature and the 'creative demon'. The relationship is highly nonlinear, complex and dynamic, evolving over time and driven by both external and internal stimuli and factors such as firm strategy, structure, and performance as well as top-down policies and bottom-up initiatives that act as enablers, catalysts, and accelerators for creativity and innovation that leads to competitiveness" (Carayannis and Gonzalez 2003, p. 593).

level we should expect homogeneity. This understanding we can paraphrase with the term "Mode 3" Knowledge Production System.

Embedding concepts of knowledge creation, diffusion, and use in the context of general systems theory could prove mutually beneficial and enriching for systems theory as well as knowledge-related fields of study, as this could:

(a) Reveal for systems theory a new and important field of application and
(b) At the same time, provide a better conceptual framework for understanding knowledge-based and knowledge-driven events and processes in the economy, and hence reveal opportunities for optimizing public sector policies and private sector practices

Thus, the major purposes of this chapter could be paraphrased as follows:

(a) *Adding to the theories and concepts of knowledge* further discursive inputs, such as suggesting a linkage of systems theory and the understanding of knowledge, emphasizing multilevel systems of knowledge and innovation, summarized also under the term of *"Mode 3" Knowledge Production Systems Approach to knowledge creation, diffusion, and use* that we discuss below.
(b) This diversified and conceptually pluralized understanding should *support practical and application-oriented decision-making with regard to knowledge, knowledge optimization, and the leveraging of knowledge for other purposes*, such as economic performance: knowledge-based decision-making has ramifications for knowledge management of firms (global multinational corporations) and universities *as well as* for public policy (knowledge policy and innovation policy).
(c) The *exploration, identification, and understanding of the key triggers, drivers, catalysts, and accelerators of high quality and quantity (continuous as well as discontinuous* and *reinforcing as well as disruptive) innovation and sustainable entrepreneurship* (financially and environmentally, see the work by the authors on the **Quintuple Innovation Helix**, in Carayannis and Campbell 2010, pp. 58–63) that serve as the foundations of robust competitiveness within the operational framework of *Open Innovation Diplomacy* (Carayannis and Campbell 2011) and **Diaspora Entrepreneurship and Innovation Networks** (Carayannis and Campbell 2011).

Since the 2009 Carayannis and Campbell article in the IJTM, we have seen several instances of pilot policy adoptions as well as implementations of the Mode 3 knowledge production system as well as the Quadruple Innovation Helix concept such as several EU FP7 RFPs and related projects implementations (such as in the Nordic and Baltic countries) as well as the integration of these concepts in the EU Innovation Union 2020 white paper (October 2010) to a significant degree. These events have reinforced and clarified our thinking and further research into the concept now taking place at different levels and diverse contexts.

1.2   Definition of Terms

1.2.1   Diplomacy

The art and practice of conducting negotiations between nations.
   A skill in handling affairs without arousing hostility.
   http://www.merriam-webster.com/dictionary/diplomacy.
   "**Diplomacy** is the art and practice of conducting negotiations between representatives of groups or states. It usually refers to international diplomacy, the conduct of international relations[1] through the intercession of professional diplomats with regard to issues of peace-making, trade, war, economics, culture, environment and human rights. International treaties are usually negotiated by diplomats prior to endorsement by national politicians. In an informal or social sense, diplomacy is the employment of tact to gain strategic advantage or to find mutually acceptable solutions to a common challenge, one set of tools being the phrasing of statements in a non-confrontational, or polite manner."
   http://en.wikipedia.org/wiki/Diplomacy.

Science Diplomacy

"What is 'Science Diplomacy'? Science Diplomacy (SD) is the exchange of Science and Technology across borders. A valuable resource and little understood tool of awareness, understanding, and capacity building, its power is not widely known or considered often enough."
   http://mountainrunner.us/2007/04/science_diplomacy.html.

Cultural Diplomacy

Cultural diplomacy specifies a form of diplomacy that carries a set of prescriptions which are material to its effectual practice; these prescriptions include the unequivocal recognition and understanding of foreign cultural dynamics and observance of the tenets that govern basic dialogue.
   Milton C. Cummings Jr. draws out the meaning of these cultural dynamics in his description of cultural diplomacy as "… the exchange of ideas, information, art, lifestyles, values systems, traditions, beliefs and other aspects of cultures…."
   http://en.wikipedia.org/wiki/Cultural_diplomacy.

Economic Diplomacy

Berridge and James (2003) state that "economic diplomacy is concerned with economic policy questions, including the work of delegations to conferences sponsored by bodies such as the WTO" and include "diplomacy which employs economic resources, either as rewards or sanctions, in pursuit of a particular foreign policy objective" also as a part of the definition.

Rana (2007) defines economic diplomacy as "the process through which countries tackle the outside world, to maximize their national gain in all the fields of activity including trade, investment and other forms of economically beneficial exchanges, where they enjoy comparative advantage.; it has bilateral, regional and multilateral dimensions, each of which is important."

http://en.wikipedia.org/wiki/Economic_diplomacy.

Innovation Diplomacy

"Science, despite its international characteristics, is no substitute for effective diplomacy. Any more than diplomatic initiatives necessarily lead to good science. These seem to have been the broad conclusions to emerge from a 3-day meeting at Wilton Park in Sussex, UK, organized by the British Foreign Office and the Royal Society, and attended by scientists, government officials, and politicians from 17 countries around the world. The definition of science diplomacy varied widely among participants. Some saw it as a subcategory of "public diplomacy," or what US diplomats have recently been promoting as "soft power" ("the carrot rather than the stick approach," as a participant described it).

Others preferred to see it as a core element of the broader concept of "innovation diplomacy," covering the politics of engagement in the familiar fields of international scientific exchange and technology transfer, but raising these to a higher level as a diplomatic objective."

http://scidevnet.wordpress.com/category/science-diplomacy-conference-2010/.

"Science and innovation together have a role that can be used to promote global equality and sustainable development," Seabra da Cruz said. He pointed out how Brazil's surging capacity in science and technology has provided a new channel for establishing relations with other countries, particularly emerging economies such as China and India, and those in other parts of the developing world:

"The big challenge to us and other emerging economies is to find ways of using scientific knowledge to enhance our competitiveness and create a new international division of labor. Without linking scientific knowledge to innovation policy, it is impossible to have sustainable development." As an example of innovation diplomacy in action, he pointed to how technical knowledge can be exchanged between countries about the best ways of using cheap, sustainable sources of energy—as Brazil is doing with its experience in biofuels—helping to improve relations between the providers of such knowledge and those that receive it. "This is an example of where we can exchange information about best social and innovation practices—which are all likely to involve science to a greater or lesser degree—and also provide an immediate and relatively easy way of making innovation work for diplomacy." He admitted that, as with science diplomacy, innovation diplomacy presents a number of challenges. Diplomats need to be well informed on innovation-related issues, embassies need to develop "observatories" that monitor the innovation landscape of the countries in which they are based, and ways need to be found to engage a country's scientific and technological "diaspora."

More specifically, Innovation Diplomacy leverages Entrepreneurship and Innovation as key drivers, catalysts, and accelerators of economic development and envisions in particular the development of efforts and initiatives along the following axes concerning in particular the socio–economic condition and dynamics in Greece currently:

1. *Re-engineer mindsets, attitudes, and behaviors* to help people—and especially the younger ones—realize the true nature and potential of innovation and entrepreneurship as a way of life and the most powerful lever for and pathway to sustainable growth and prosperity with positive spill-over effects staunching the braindrain, reduced cynicism, and increased optimism and trust in the future and each other, reduced criminality and social unrest, higher assimilation of migrant groups, etc.
2. *Engage in sustained, succinct, and effective dialog with stakeholders and policy makers within the involved countries* to pursue the reform and as needed re-invention of institutions, policies, and practices that can make flourish entrepreneurship and innovation in areas such as related laws, rules and regulations, higher education, public and private Research and Development, civil society movements and nonGovernmental organizations, etc.
3. *Identify, network, and engage purposefully and effectively with the Diaspora professional and social networks around the world* to trigger, catalyze, and accelerate their involvement and intervention in a focused and structured manner to help with goals 1 and 2 above as well as help establish, fund, and manage entrepreneurship and innovation promoting and supporting initiatives and institutions such as business plan competitions, angel, and other risk capital financing of new ventures, mentoring of, and partnering with said ventures to ensure their survival, growth, and success both within a given country and in the global markets. Of particular interest and importance would be communities of practice and interest among the Diaspora Entrepreneurship and Innovation Networks.

To fully leverage the potential of systems (and systems theory) one should also demonstrate, how a system design can be brought in line with other available concepts, such as innovation networks and knowledge clusters. With regard to clusters, at least three types of clusters can be listed:

1. *Geographic (spatial) clusters*: In that understanding, a cluster represents a certain geographic, spatial configuration, either tied to a location or a larger region. Geographic, spatial proximity, for example for the exchange of tacit knowledge, is considered as crucial. While "local" clearly represents a subnational entity, a "region" could be either subnational or transnational.
2. *Sectoral clusters*: This cluster approach is carried by the understanding that different industrial or business sectors develop specific profiles with regard to knowledge production, diffusion, and use. One could even add that sectoral clusters even support the advancement of particular "knowledge cultures." In innovation research, the term "innovation culture" already is being acknowledged (Kuhlmann 2001, p. 958).

3. *Knowledge clusters*: Here, a cluster represents a specific configuration of knowledge, and possibly also of knowledge types. However, in geographic (spatial) and sectoral terms, a knowledge cluster is not predetermined. In fact, a knowledge cluster can cross-cut different geographic locations and sectors, thus operating globally and locally (across a whole multilevel spectrum). Crucial for a knowledge is, if it expresses an innovative capability, for example produces knowledge that excels (knowledge-based) economic performance. A knowledge cluster, furthermore, may even include more than one geographic and/or sectoral clusters.

Networks emphasize *interaction, connectivity, and mutual complementarity and reinforcement*. Networks, for example, can be regarded as the internal configuration that ties together and determines a cluster. Networks also can express the relationship between different clusters. *Innovation networks and knowledge clusters thus resemble a matrix*, indicating the interactive complexity of knowledge and innovation. Should the (proposed) conceptual flexibility of systems (and systems theory) be fully leveraged, it appears important to demonstrate how systems relate conceptually to knowledge clusters and innovation networks, as they are key in understanding the nature and dynamics of knowledge stocks and flows. What we suggest is to link the two basic components (attributes) of systems ("elements/parts" and "rationale/self-rationale"; Campbell 2001, p. 426) with clusters and networks (Carayannis and Campbell 2006a, pp. 9–10). What results is a formation of two pairs of theoretical equivalents (see Fig. 1)[11]:

1. *Elements and clusters*: The elements (parts) of a system can be regarded as an equivalent to clusters (knowledge clusters).
2. *Rationale and networks*: The rationale (self-rationale) of a system can be understood as an equivalent to networks (innovation networks).

The rationale of a system holds together the system elements and expresses the relationship between different systems. It could be argued that, at least partially, this rationale manifests itself ("moves through") as networks. At the same time, elements of a system might also manifest themselves as clusters. Perhaps, networks could be affiliated with the functions of a system, and clusters with the structures of systems. This would help indicating to us, should we be interested in searching for structures and functions of knowledge and innovation systems, what exactly to look for. This, obviously, does not imply to claim that structures and functions of knowledge (innovation) systems only fall into the conceptual boxes of "clusters" and "networks." However, clusters and networks should be regarded as crucial subsets for the elements and rationales of systems.[12]

---

[11] Of course there may also be *systems of clusters and networks* or *clusters and networks of systems*.

[12] For an example of an interesting analysis of the applicability of systems or networks for the field of research policy, see Sylvia Kritzinger et al. (2006).

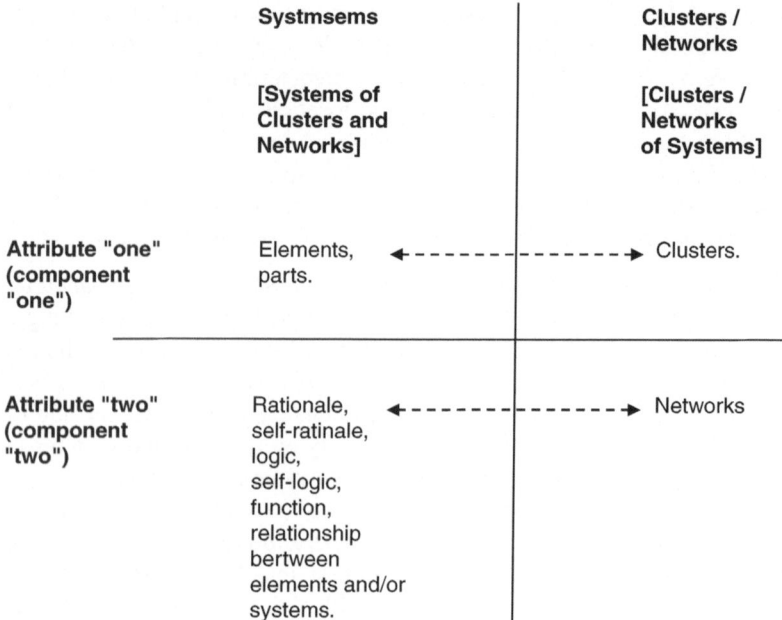

Fig. 1 Theoretical equivalents between conceptual attributes of systems and clusters/networks. Source: Authors' own conceptualization based on Carayannis and Campbell (2009, p. 204)

This equation formula (between elements/clusters and rationales/networks) might need further conceptual and theoretical development. But it lays open a convincing route for better understanding knowledge and innovation, through tying together two strong conceptual traditions (systems theory with clusters and knowledge). A further ramification of networks, as we will demonstrate later on, could also imply to understand (at least the large-scale) knowledge strategies as complex network configurations.

As a new input for discussion, we wish to introduce the concept of *the* "Mode 3" *knowledge creation, diffusion, and use system,* and we define below the essential elements or building blocks of "Mode 3." The notion "Mode 3" was coined by Carayannis (late fall of 2003), and was as a concept jointly developed by Carayannis and Campbell (2006a).

In the following, we list some of the key definitions, which refer to "Mode 3" and associated concepts (see also Carayannis and Campbell 2006c, 2009).

- *The "MODE 3" Systems Approach for Knowledge Creation, Diffusion, and Use:"Mode 3" is a multilateral, multinodal, multimodal, and multilevel systems approach to the conceptualization, design, and management of real and virtual, "knowledge-stock," and "knowledge-flow," modalities that catalyze, accelerate, and support the creation, diffusion, sharing, absorption, and use of co-specialized knowledge assets. "Mode 3" is based on a system-theoretic perspective of socio-economic, political, technological, and cultural trends and conditions that shape the co-evolution of knowledge with the "knowledge-based and knowledge-driven, gloCal economy and society."*[13]

- ***Innovation Networks***: *Innovation Networks*[14] *are real and virtual infrastructures and infratechnologies that serve to nurture creativity, trigger invention, and catalyze innovation in a public and/or private domain context (for instance, Government-University-Industry Public-Private Research and Technology Development Co-opetitive Partnerships*[15,16]*).*
- ***Knowledge Clusters***: *Knowledge Clusters are agglomerations of co-specialized, mutually complementary and reinforcing knowledge assets in the form of "knowledge stocks" and "knowledge flows" that exhibit self-organizing, learning-driven, dynamically adaptive competences and trends in the context of an open systems perspective.*
- ***Twenty-first Century Fractal Research, Education and Innovation Ecosystem (FREIE)***: *A twenty-first Century FREIE is a multilevel, multimodal, multinodal, and multi-agent system of systems. The constituent systems consist of innovation meta-networks (networks of innovation networks and knowledge clusters) and knowledge meta-clusters (clusters of innovation networks and knowledge clusters) as building blocks and organized in a self-referential or chaotic*[17]

---

[13] Carayannis and Zedwitz (2005).

[14] Networking is important for understanding the dynamics of advanced and knowledge-based societies. Networking links together different modes of knowledge production and knowledge use, and also connects (subnationally, nationally, and transnationally) different sectors or systems of society. Systems theory, as presented here, is flexible enough for integrating and reconciling systems and networks, thus creating conceptual synergies.

[15] Carayannis and Alexander (2004).

[16] Carayannis and Alexander (1999a).

[17] Carayannis (2001, pp. 169–170) discusses chaos theory and fractals in connection to technological learning and knowledge and innovation system architectures: "Chaos theory is a close relative of catastrophe theory, but has shown more potential in both explaining and predicting unstable non-linearities, thanks to the concept of self-similarity or fractals [*patterns within patterns*] and the chaotic behavior of attractors (Mandelbrot) as well as the significance assigned to the role that initial conditions play as determinants of the future evolution of a non-linear system (Gleick 1987). There is a strong affinity with strategic incrementalism, viewed as a third-order (triple-layered), feedback-driven system that can exhibit instability in any given state as a result of the operational, tactical, and strategic technological learning … that takes place within the organization in question."

fractal[18] *(Gleick* 1987*) knowledge and innovation architecture (Carayannis* 2001*), which in turn constitute agglomerations of human, social, intellectual, and financial capital stocks and flows as well as cultural and technological artifacts and modalities, continually co-evolving, co-specializing, and co-opeting. These innovation networks and knowledge clusters also form, re-form, and dissolve within diverse institutional, political, technological, and socio-economic domains including Government, University, Industry, and Nongovernmental Organizations and involving Information and Communication Technologies, Biotechnologies, Advanced Materials, Nanotechnologies, and Next Generation Energy Technologies.*

A fractal innovation ecosystem (a special case being a FREIE—see the article by Carayannis and Campbell 2011) is an agglomeration of resources that act, interact, and evolve under regimes of co-opetition, co-specialization, and co-evolution in pursuit of higher levels of effectiveness and efficiency in resource creation, allocation, appropriation, and use. As a result of these processes and dynamics, the architecture and topology of said ecosystem materialize in a manner that emulates networks of innovation and clusters of knowledge (for reasons of proximity, affinity, density, and the like). Overall, these structures tend to become self-similar, hence the fractal nature of the architecture and topology, as this seems to afford higher efficacy levels and moreover, higher likelihood for strategic knowledge serendipity and arbitrage events (Carayannis 2008) which is part of our ongoing research.

1.3   Mode 3, Quadruple Helix, Quintuple Helix, Democracy of Knowledge, Schumpeter's Creative Destruction, and the Co-evolution of Different Knowledge Modes

Per the comments above, the Triple, Quadruple, and Quintuple Innovation Helices are in effect topologically equivalent modalities with varying degrees of complexity and dimensionality (moving from three to four to five degrees—Government,

---

[18] "A **fractal** is a geometric object which is rough or irregular on all scales of length, and so which appears to be 'broken up' in a radical way. Some of the best examples can be divided into parts, each of which is similar to the original object. Fractals are said to possess infinite detail, and some of them have a self-similar structure that occurs at different levels of magnification. In many cases, a fractal can be generated by a repeating pattern, in a typically recursive or iterative process. The term *fractal* was coined in 1975 by Benoît Mandelbrot, from the Latin *fractus* or 'broken'. Before Mandelbrot coined his term, the common name for such structures (the Koch snowflake, for example) was *monster curve*. Fractals of many kinds were originally studied as mathematical objects. *Fractal geometry* is the branch of mathematics which studies the properties and behavior of fractals. It describes many situations which cannot be explained easily by classical geometry, and has often been applied in science, technology, and computer-generated art. The conceptual roots of fractals can be traced to attempts to measure the size of objects for which traditional definitions based on Euclidean geometry or calculus fail." (http://en.wikipedia.org/wiki/Fractal).

University, Industry, and then adding Civil Society and then adding the Environment). In this regard, they are all pillars of a fractal innovation ecosystem or more accurately and per the above comments, a FREIE.

In the following segments, we present in greater detail different aspects of advanced knowledge and innovation. Crucial for the suggested "Mode 3" approach is the idea that an advanced knowledge system may integrate different knowledge modes. Some knowledge (innovation) modes certainly will phase out and stop existing. However, what is important for the broader picture is that in fact co-evolution, co-development, and co-specialization of different knowledge modes emerge. This pluralism of knowledge modes should be regarded as essential for advanced knowledge-based societies and economies. This may point to similar features of advanced knowledge and advanced democracy. We could state that competitiveness and sustainability of the gloCal knowledge economy and society increasingly depend on the elasticity and flexibility of promoting a co-evolution and by this also a cross-integration of different knowledge (innovation) modes. This heterogeneity of knowledge modes should create hybrid synergies and additionalities.

The "Triple Helix" model of knowledge, developed by Henry Etzkowitz and Loet Leydesdorff (2000, pp. 111–112), stresses three "helices" that intertwine and by this generate a national innovation system: academia/universities, industry, and state/government. Etzkowitz and Leydesdorff are inclined of speaking of "university-industry-government relations" and networks, also placing a particular emphasis on "tri-lateral networks and hybrid organizations," where those helices overlap. In extension of the Triple Helix model we suggest a "Quadruple Helix" model (see Fig. 2). *Quadruple Helix, in this context, means to add to the above stated helices a "fourth helix" that we identify twofold, as the "media-based and culture-based public" as well as the "civil society"* (see, furthermore, Carayannis and Campbell 2009, pp. 206–207; Danilda et al. 2009; Lindberg et al. 2012; Colapinto and Porlezza 2012). This should emphasize that a broader understanding of knowledge production and innovation application requires that also the public becomes more integrated into advanced innovation systems. The public uses and applies knowledge, so public users are also part of the innovation system. In an advanced knowledge society and knowledge economy, knowledge flows out into all spheres of society. When we speak of the "public" in context of the Quadruple Helix, we mean in more particular: the media-based and culture-based public and civil society. But also other aspects are being addressed as well: culture (cultures) and innovation culture (innovation cultures)[19]; *the knowledge of culture and the culture of knowledge* (Carayannis 2012); values and life styles; multiculturalism, multiculture, and creativity; media; arts and arts universities; and multilevel innovation systems (local, national, and global), with universities of the sciences, but also universities of the arts. These diverse and heterogeneous settings of culture should help fostering creativity, which is so necessary and essential for creating and producing new knowledge and new innovations. *"We can also call this the creativity of knowledge creation"*

---

[19] On "innovation culture," see also: Kuhlmann 2001, pp. 954, 958 and 962.

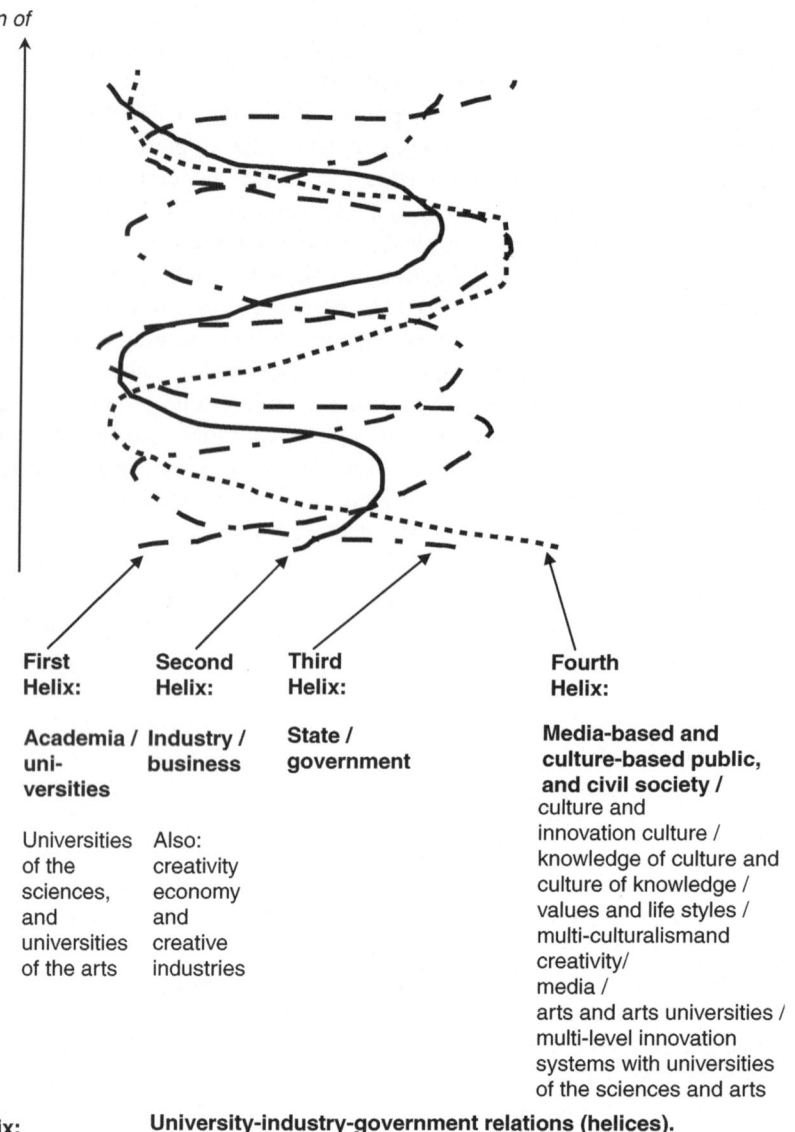

Direction of time

First Helix:

Academia / universities

Universities of the sciences, and universities of the arts

Second Helix:

Industry / business

Also: creativity economy and creative industries

Third Helix:

State / government

Fourth Helix:

Media-based and culture-based public, and civil society / culture and innovation culture / knowledge of culture and culture of knowledge / values and life styles / multi-culturalismand creativity/ media / arts and arts universities / multi-level innovation systems with universities of the sciences and arts

Triple Helix: **University-industry-government relations (helices).**

Quadruple Helix, Fourth Helix: **"Media-based and culture-based public", and "civil society".**

Fig. 2 The conceptualization of the "Quadruple Helix" innovation system. Source: Authors' own conceptualization based on Etzkowitz and Leydesdorff (2000, p. 112), Carayannis and Campbell (2009, p. 207, 2010, p. 62) and Danilda et al. (2009)

(Carayannis and Campbell 2010, p. 48). In organizational and institutional terms, this encourages developing "Creative Knowledge Environments." Hemlinet al. (2004, p. 1) define such contexts in the following way: "Creative knowledge environments (CKEs) are those environments, contexts and surroundings the characteristics of which are such that they exert a positive influence on human beings engaged in creative work aiming to produce new knowledge or innovations, whether they work individually or in teams, within a single organization or in collaboration with others" (see also Resetarits and Resetarits-Tincul 2012). Richard Florida (2004) coined the notion of the "creative class" (a term, coined by Richard Florida 2004). Plausibility for the explanatory potential of such a fourth helix is that culture and values, on the one hand, and the way how "public reality" is being constructed and communicated by the media, on the other hand, influence every national and every multilevel innovation system. The proper "innovation culture" is here key for promoting an advanced knowledge-based economy. Through public discourses, transported through and interpreted by the media, are crucial for a society to assign top priorities to innovation and knowledge (research, technology, and education).

The creative industries are part of an economy, in context of the Quadruple Helix. It is reasonable, however, not only to speak of the creative industries, but also to envision more comprehensively a "creativity economy," where creativity is relevant for all sectors of the economy as well as all sectors of society. An advanced knowledge economy is a knowledge economy, innovation economy, and a creativity economy at the same time. The more mature and advanced a knowledge economy, innovation economy, and knowledge society are, the more creativity is being demanded. As Dubina et al. (2012) state: "The more advanced and mature a knowledge economy (creativity economy) and knowledge society (creativity society) are, the more knowledge, innovation and creativity can be absorbed and are even being demanded for further progress. *The creativity economy creatively interrelates technological innovations with social innovations*" (see Fig. 3).

In the multilevel innovations systems, which are being carried and driven by advanced knowledge production in context of the Quadruple Helix innovation model, research activities of the universities of the sciences (natural sciences, life sciences, social sciences, and humanities) are essential. However, what counts here, are not only the sciences, but also the arts. The sciences are a manifestation of knowledge, but also the arts, at least partially, can be understood as a manifestation of knowledge. In context of higher education and the universities, we are often inclined to speak of "scientific research." But there exist also important forms of "artistic research." Artistic research, in fact, represents an innovative conceptualization of a new form of art creation and art practice, possibly also a new form of knowledge creation. "'Artistic research' is a new practice in the arts in which artists themselves act as researchers and present their findings in the form of artwork. This practice is firmly established at European universities but has so far provoked little public response. What distinguishes artistic research from 'mere' art, and what contributions can it make to the art world?" (Caduff et al. 2010, cover page; see also McNiff 1998, 2008, and, furthermore, Ritterman et al. 2011). Artistic research and research in the arts can engage in interdisciplinary and transdisciplinary network

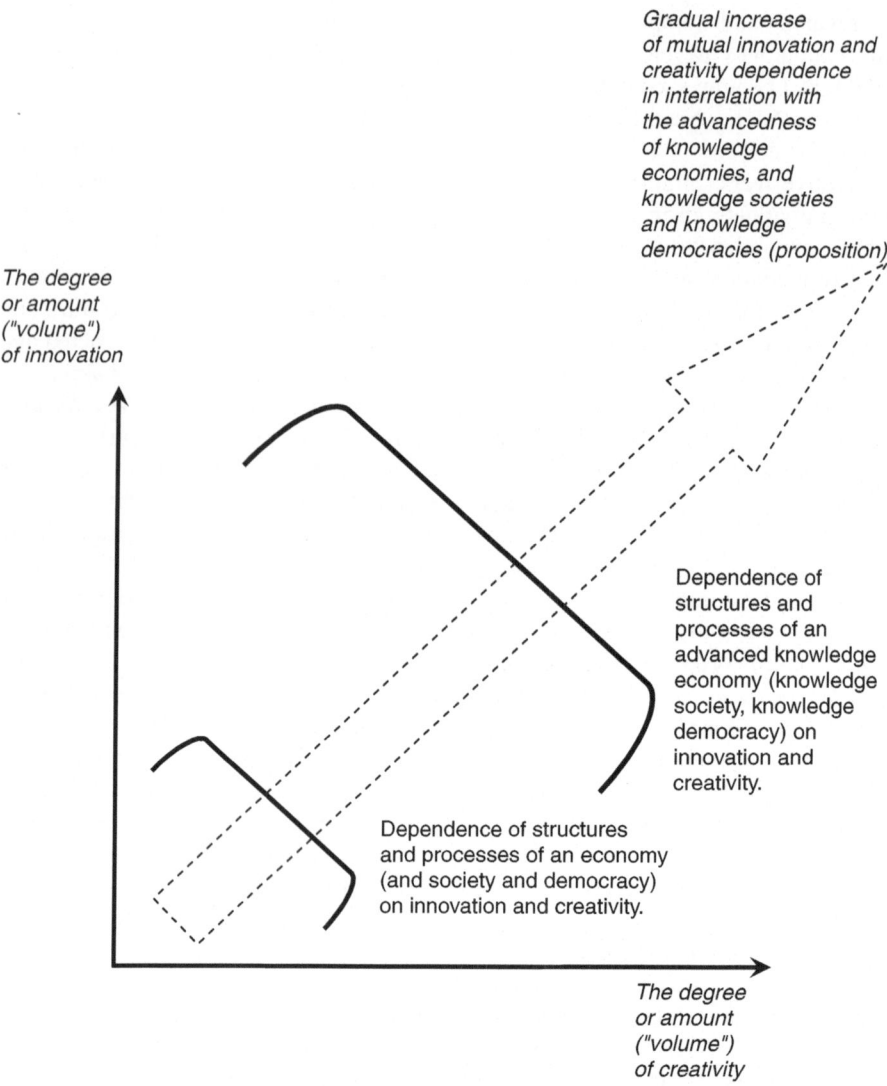

Fig. 3 The increasing cross-interrelation of innovation and creativity in advanced knowledge economies, knowledge societies and knowledge democracies. Source: Authors' own conceptualization based on Dubina et al. (2012)

arrangements with research in the sciences. Artistic research and universities of the arts should be regarded as being of a crucial importance for multilevel innovation systems in advanced knowledge economies that are also creativity economies (see also Yau 2012). *Artistic research, research in the arts and arts universities, in hybrid, pluralized, and heterogeneous combinations with universities of the sciences and*

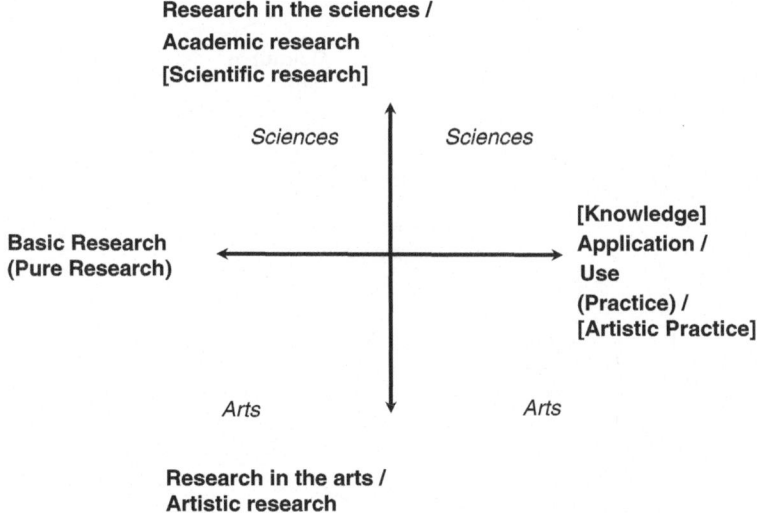

Fig. 4 Research and knowledge application in the sciences and arts. Source: Authors' own conceptualization

research in the sciences, add to the creativity of new knowledge production and new innovations. In the sciences, there is often the understanding of a spectrum from basic (pure) research to applied research. Also for the arts, one may propose a spectrum of (pure) basic artistic research to the (applied) practice of arts (see Fig. 4).[20]

The Triple Helix may be regarded as a "core model" for innovation, resulting from interactions in knowledge production referring to universities (higher education), industries (economy), and governments (multilevel). The Triple Helix is being contextualized by the broader innovation model of the Quadruple Helix, which is blending in features of the public, for example civil society and the media-based and culture-based public. The Quintuple Helix innovation model, finally, contextualizes the Quadruple Helix (and Triple Helix). The Quintuple Helix brings in the perspective of the natural environments of society and the economy for knowledge production and the innovation systems. "For the purpose of further discussion and analysis we lastly want to propose and introduce the five-helix model of the 'Quintuple Helix', where the environment or the natural environments represent the fifth helix" (Carayannis and Campbell 2010, p. 61). Furthermore: "The Quintuple Helix can be proposed as a framework for transdisciplinary (and interdisciplinary) analysis of sustainable development and social ecology" (Carayannis and Campbell 2010, p. 62). A sustainable balance between the paths of development of society and the

---

[20] Figure 4 should be seen here as a suggestion, as an input for discussion. The conceptual feasibility of Fig. 4 still would have to be tested.

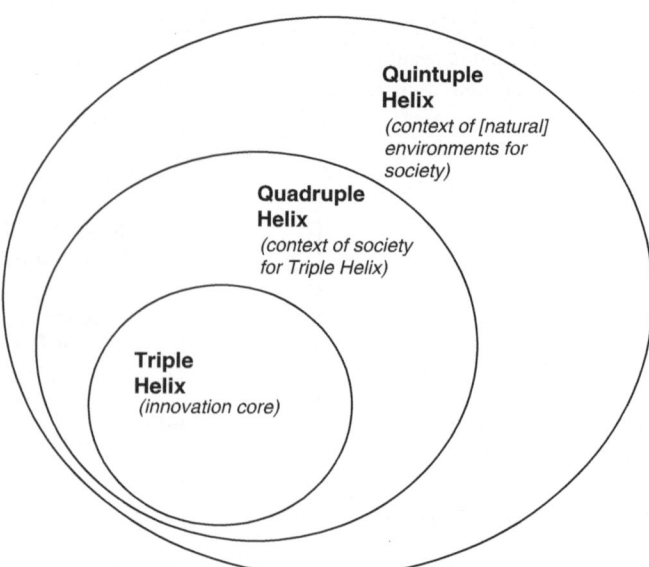

Fig. 5 Society as context for Triple Helix innovation systems, natural environments as context for Quadruple Helix innovation systems. Source: Authors' own conceptualization based on Etzkowitz and Leydesdorff (2000, p. 112), Carayannis and Campbell (2009, p. 207, 2010, p. 62), and Danilda et al. (2009)

economy, with their natural environments, is essential for the further progress of human civilizations. The Quintuple Helix, however, also emphasizes that the natural environments should be conceptualized as drivers for the further advancing of knowledge production and innovation systems. Thus, the Quintuple Helix model appears to be compatible with the interests, also analytical interests of "social ecology" and "sustainable development" (on social ecology, see Fischer-Kowalski and Haberl 2007; on "sustainable development" see Winiwarter and Knoll 2007, pp. 305, 306–307). *With the transdisciplinary application of interdisciplinary[21] knowledge[22] (also of the sciences and arts) the Quintuple Helix wants to create and support a mid-term and long-term sustainable development of society, the economy, and democracy that is sensitive for social ecology as well as social-ecologically-friendly.*[23] The Quadruple Helix contextualizes the Triple Helix, and the Quintuple Helix contextualizes the Quadruple Helix (see Fig. 5). Depending on the interests and the analytical interests, it could be equally appropriate to frame a research question in reference to the Triple Helix, Quadruple Helix, or Quintuple Helix innovation

---

[21] On interdisciplinarity ("*Interdisziplinarität*"), see Markus Arnold (2009, pp. 65–97).

[22] For interesting examples of integrating and analytically combining research in the fields and disciplines of the social sciences and natural sciences, see: Gottweis 1998; Hindmarsh and Prainsack 2010; Prainsack and Wolinsky 2010.

[23] On potential epistemic (epistemological) implications of the Quintuple Helix model for society-nature interactions and social ecology, see Campbell and Campbell (2011, pp. 15–16, 23–27).

models. However, even when an analysis or assessment is being carried out in a Triple Helix framework, also, at one point, the contexts of Quadruple Helix and Quintuple Helix should be taken into consideration. The knowledge and innovation perspectives of Quadruple and Quintuple Helix are broader; thus they add crucially to the prospects and opportunities of a sustainable problem-solving. The more advanced knowledge societies and knowledge economies are progressing, the more there is a need to shift the attention to broader innovation models (see Fig. 6).[24]

Figure 7 displays visually from which conceptual perspectives the co-evolution and cross-integration of different knowledge modes could be approached. Mode 3 emphasizes the additionality and surplus effect of a *co-evolution of a pluralism of knowledge and innovation modes*. Quadruple Helix refers to structures and processes of the gloCal (global and local) knowledge economy and society; Quintuple Helix also brings in the perspective of the natural environments (social ecology). Furthermore, the "Innovation Ecosystem," combining and integrating social and natural systems and environments, stresses the importance of a pluralism of a diversity of agents, actors, and organizations: universities (universities of the sciences and arts), small and medium-sized enterprises, and major corporations, arranged along the matrix of fluid and heterogeneous innovation networks and knowledge clusters. This all may result in a *Democracy of Knowledge*, driven by a pluralism of knowledge and innovation and by a pluralism of paradigms of knowledge modes. *The democracy of knowledge, as a concept and metaphor, is being carried by the understanding that there operates (at least potentially) a co-evolution between processes of advancing democracy and processes of advancing knowledge and innovation.* Here the knowledge democracy and knowledge economy meet and overlap. *Between processes and structures of advanced knowledge democracy, knowledge society, and knowledge economy, there is a certain congruence* (Carayannis and Campbell 2010, pp. 54–58, 60–61). Concepts of democracy (moving from electoral to liberal and high-quality democracies), and of knowledge and innovation (e.g., re-focusing from Triple Helix to Quadruple and Quintuple Helices), are becoming broader and increase their complexity considerably. *Political pluralism in democracy cross-refers to creativity-encouraging heterogeneity and diversity of different forms, modes, and paradigms of knowledge and innovation.*[25] In "The Republic of Science," Michael Polanyi (1962, p. 54) expressed already some similar ideas: "My title is intended to suggest that the community of scientists is organized in a way which resembles certain features of a body politic and works according to economic

---

[24] Loet Leydesdorff (2012) launched the interesting intellectual experiment of engaging in theorizing on "*N*-Tuple of Helices" of innovation systems, introducing a multi-dimensional view perspective. In abstract terms, one may always refer to a structure (matrix structure) of *N*-Tuple Helices. *However, in the models of innovation, being presented here, we propose concrete characteristics and properties of the Quadruple and Quintuple Helix so to transform these into meaningful tools for empirical analysis and application.* There are also epistemic (epistemological) qualities of the Quadruple and Quintuple Helixes innovations systems.

[25] This, of course, also challenges our external and internal governance models of higher education. For an overview on governance approaches in higher education, see Ferlie et al. (2008, 2009). See also Biegelbauer (2010). On structures and changes of universities, see also Krücken (2003a, 2003b), and Krücken et al. (2007).

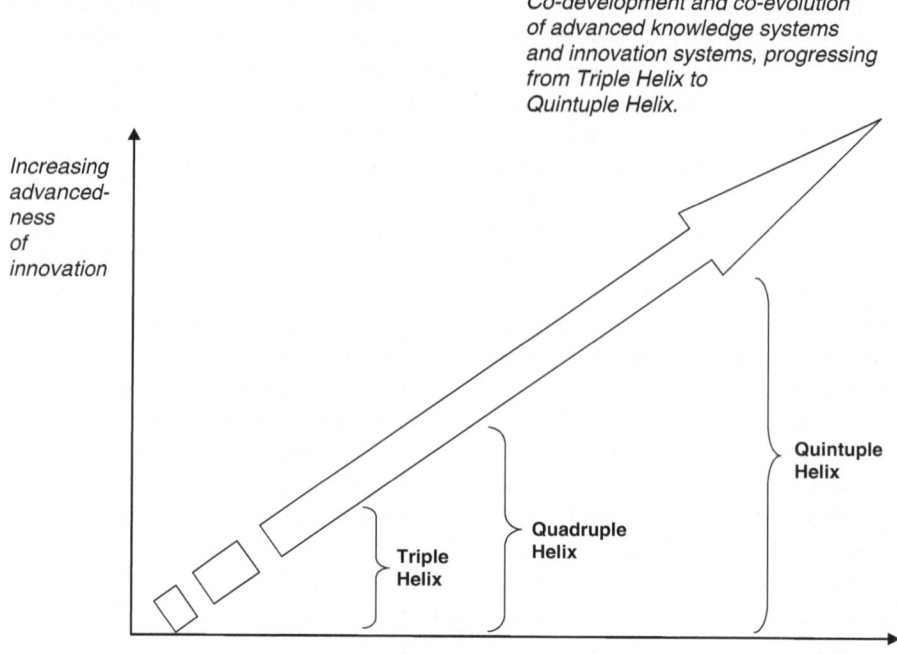

Fig. 6 The co-development and co-evolution of advanced knowledge production and andvanced innovation systems. Source: Authors' own conceptualization based on Etzkowitz and Leydesdorff (2000, p. 112), Carayannis and Campbell (2009, p. 207, 2010, p. 62), and Danilda et al. (2009)

principles similar to those by which the production of material goods is regulated." We suggest here that the *Democracy of Knowledge* contextualizes the *Republic of Science* in an already broader perspective.

In the "Frascati Manual," the Organization for Economic Co-operation and Development (OECD 1994, p. 29) distinguishes between the following activity categories of research (R&D, research and experimental development): basic research, applied research, and experimental development. Basic research represents a primary competence of university research, whereas business R&D focuses heavily on experimental development. Assessed empirically for the United States, one of the

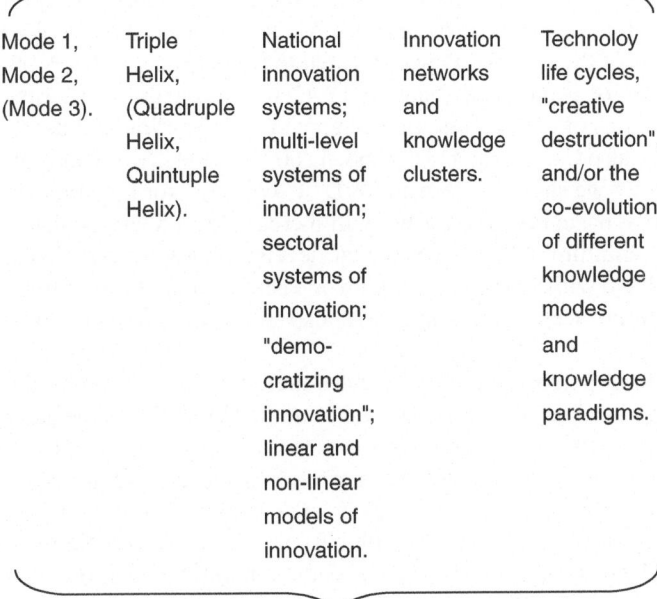

**Knowledge production
integration of: Mode 3**

| Mode 1,
Mode 2,
(Mode 3). | Triple
Helix,
(Quadruple
Helix,
Quintuple
Helix). | National
innovation
systems;
multi-level
systems of
innovation;
sectoral
systems of
innovation;
"demo-
cratizing
innovation";
linear and
non-linear
models of
innovation. | Innovation
networks
and
knowledge
clusters. | Technoloy
life cycles,
"creative
destruction",
and/or the
co-evolution
of different
knowledge
modes
and
knowledge
paradigms. |

**Knowledge
innovation
integration of:
Quadruple
Helix

(context of
society,
social
environments)**

**Knowledge innovation
integration of:**
**Innovation Ecosystem, social and
natural environments, Quintuple Helix**

**"Democracy of Knowledge"
(contextualizing the
"Republic of Science"):**
Democracy of Knowledge of knowledge creation, diffusion and use;
co-evolution (congruence) of political pluralism in
advanced democracy and knowledge (innovation)
heterogeneity in advanced economy and society.

Fig. 7 Knowledge creation, diffusion and use in a Democracy of Knowledge. Source: Authors'
own conceptualization based on Carayannis and Campbell (2009, p. 208); see also Von Hippel
(2005) and Polanyi (1962)

globally leading national innovation systems, with regard to the financial volume of R&D resources the experimental development ranks first, applied research second, and basic research third (OECD 2006; National Science Board 2010, Chap. 4, pp. 8–16). Interesting, however, is the dynamic momentum, when observed for a longer period of time. Basic research, in the U.S.A., grew faster than applied research. In 1981, 13.4% of the U.S. R&D was devoted to basic research. By 2008, basic research increased its percentage share to 17.47%. During the same time period the percentage shares of applied research stagnated and experimental development even declined (see also Carayannis and Campbell 2009, pp. 209–210). This links up to the question, whether we should expect an R&D "U-curving" for U.S. innovation system, implying that basic research further will increase its percentage shares of the overall R&D expenditure while experimental development may slide back. This would go hand-in-hand with an importance gain of basic research. Furthermore, would such a potential future scenario for the U.S.A. also spill over to other national innovation systems?

Assessed in a long-term perspective (1953–2008), there has been a substantial shift in the financing and funding of the national R&D in the U.S.A. Until the early 1970s, the federal government was the most important funding source for R&D. After that business moved up to become the primary funding source, and gradually increased its dominance since then. During the 1970s, the funding base of national R&D in the U.S.A. converted from primarily public to primarily private (National Science Board 2010, Chap. 4, pp. 11, 14). This feeds general expectations that mature and advanced national R&D systems are being funded and performed, first of all, by the economy (the business enterprise sector). In less advanced R&D systems, the role of business is less important, in relative terms. However, and this appears to be a crucial argument here: this important gain of the economy does not imply that basic or applied research is becoming less important. What seems to count then is the basic and applied research conducted by business. *Business basic research creates key opportunities to interact, cross-link, and network with the university basic research in the higher education sector, thus fostering hybrid knowledge and innovation interactions, in a linear and nonlinear fashion.*

The OECD (2002, p. 30) provides the following definition for basic research: "***Basic research*** is experimental or theoretical work undertaken primarily to acquire new knowledge of the underlying foundation of phenomena and observable facts, without any particular application or use in view." We should raise the question, whether this is still an appropriate or sufficient definition for basic research? The problem is that this definition creates a contradiction between basic research and application, but why? In the old world of a dominance of "Mode 1" for the universities this may have been a legitimate position or proposition, but in the new worlds of Mode 2 and Mode 3 of knowledge production, this general exclusion of application, for basic research, does not make sense. In the old world of knowledge production, perhaps there was a reasonable interest in a sharp line of division (boundary) between basic and applied research. Nowadays, *basic research in the context of application* has risen to new prominence and importance, and may be one of the keys for remodeling our knowledge and innovation systems. So there also appears to be a need or even a demand for a more "application-friendly" redefinition of basic research. The here suggested phrasing for a re-definition of basic research could be as follows: "***Basic research*** is experimental or theoretical work undertaken primarily to acquire new knowledge of the underlying foundation of phenomena and observable facts, *without or with* a particular application or use in view (in the long run)." Such a re-definition nicely balances the qualities of basic research with the opportunities of more simultaneously coupling basic research with application, linearly and nonlinearly. There is even a chance that the established definition of basic research, quoted above and still being used by the OECD, really underestimates the extent of basic research that already is being conducted by the economy. Is the economy (in the advanced knowledge economy) performing more *basic business research* than the conventional definitions capture and reflect? Our proposed conceptual re-definition of basic research may radically and substantially shift and transform our assessment of the patterns and behavior of advanced knowledge and innovation.

In a simple understanding, the "linear model of innovation" claims: first, there is basic university research. Later this basic research converts into applied research of intermediary organizations (university-related institutions).[26] Finally, firms pick up, and transform applied research to experimental development, which is then being introduced as commercial market applications. This linear understanding often is referred to Vannevar Bush (1945), even though Bush himself, in his famous report, neither mentions the terms "linear model of innovation" nor even the word "innovation." "Nonlinear models of innovation," on the contrary, underscore a more parallel coupling of basic research, applied research, and experimental development. Thus universities or HEIs (higher education institutions)[27] in general, university-related institutions and firms join together in variable networks and platforms for creating innovation networks and knowledge clusters. Even though there continues to be a division of labor and a functional specialization of organizations with regard to the type of R&D activity, universities, university-related institutions, and firms can perform, at the same time, basic and applied research and experimental development.

---

[26] In the German language, "university-related" would qualify as "außeruniversitär" (Campbell 2003, p. 99).

[27] Hans Pechar and Lesley Andres (2011, p. 25) carried out an analysis that compared "welfare regimes and higher education". While all OECD "…countries have experienced an unprecedented expansion in higher education during the second half of the twentieth century," they "…differ, however, with respect to the significance of education, and more specifically, higher-education policies within their overall framework of welfare policies". So Pechar and Andres (2011, p. 25) "employ the concept of the 'welfare regime' and a 'trade-off' hypothesis to understand the different national approaches to higher-education". Concerning "welfare-state regimes," Pechar and Andres refer to a typology of Gøsta Esping-Andersen. Esping-Andersen (1990, pp. 50–54) identifies three types of welfare regimes (see also Pechar and Andres 2011): *liberal welfare regimes* (for example Canada and the U.S., Australia and New Zealand, the UK); *conservative welfare regimes* (for example Austria, France, Germany, the Netherlands and Belgium, Switzerland, Italy); and *social-democratic (universal) welfare regimes* (for example Norway, Sweden, Denmark, and Finland).

Surveys about sectoral innovation in the pharmaceutical sector (McKelvey et al. 2004) and the chemical sector (Cesaroni et al. 2004) reveal how each of these industries may be characterized by complex network configurations and arrangement of a diversity of academic and firm actors. The Mode 3 Innovations Ecosystem thus represents a model of an interactive coupling of "nonlinear innovation modes": *partially, this also could mean linking together "linear innovation modes" of different degrees of maturity in the knowledge value chain or closeness to market application*, fostering the set-up of "creative knowledge environments" in organizations and institutions (see Fig. 8). We can speculate, whether this parallel integration of linearity and nonlinearity not also encourages a new approach of paralleling in our theorizing of and viewing on causality: *in epistemic (epistemological) terms, the so-called if-then relationships could be complemented by (a thinking in) "if-if" relations* (Campbell 2009, p. 123).[28] *Cross-employment* (multi-employment) may be regarded as one (organizational) strategy for realizing creative knowledge environments. Cross-employment (multi-employment) refers to a knowledge worker, employee, who is being simultaneously employed by more than one organization, possibly being located in different sectors (e.g., a higher education and a non-higher education institution, e.g., a university and a firm). *This supports the direct network-style coupling of very different organizations in knowledge production and innovation application*, expressing, therefore, what nonlinear innovation could mean in practical terms (Campbell 2011). Cross-employment makes possible "parallel careers" for individuals (knowledge workers) across a diversity of organizations and sectors, thus also a simultaneous operating in parallel in organizations with different rationales and innovation cultures.

The concept of the "entrepreneurial university" captures the need of linking more closely together university research with the R&D market activities of firms (see, e.g., Etzkowitz 2003). Mode 1 refers to a university knowledge production that focuses on basic university research that is interested in delivering comprehensive explanations of the world, structured in a "disciplinary logic," and not (per se) interested in knowledge application and innovation. Mode 2 refers to a university knowledge production that is based on the following principles: (1) "knowledge produced in the context of application"; (2) "transdisciplinarity"; (3) "heterogeneity and organizational diversity"; (4) "social accountability and reflexivity"; and (5) "quality control" (see Gibbons et al. 1994, 3–8, 167). "Mode 2" universities and "Entrepreneurial Universities" overlap, at least conceptually. *A "Mode 3" university (higher education institution, also subunit) or "Mode 3" higher education sector would be an organization or a system that operates simultaneously according to the two knowledge principles of Mode 1 and Mode 2.* Mode 3 universities seek organizational designs in trying to combine, in co-evolving and co-learning patterns, Mode 1 and Mode 2 by believing that this creates a surplus in high-quality, creative, and

---

[28] "Fortgesetzt, bezogen auf Modelle über oder von 'Kausalität', ließe sich andenken, die *Wenn-dann*-Beziehungen ('if-then relations') mit systematischen *Wenn-wenn*-Beziehungen ('if-if relations') zu ergänzen" (Campbell 2009, p. 123).

Model of linear innovation modes:

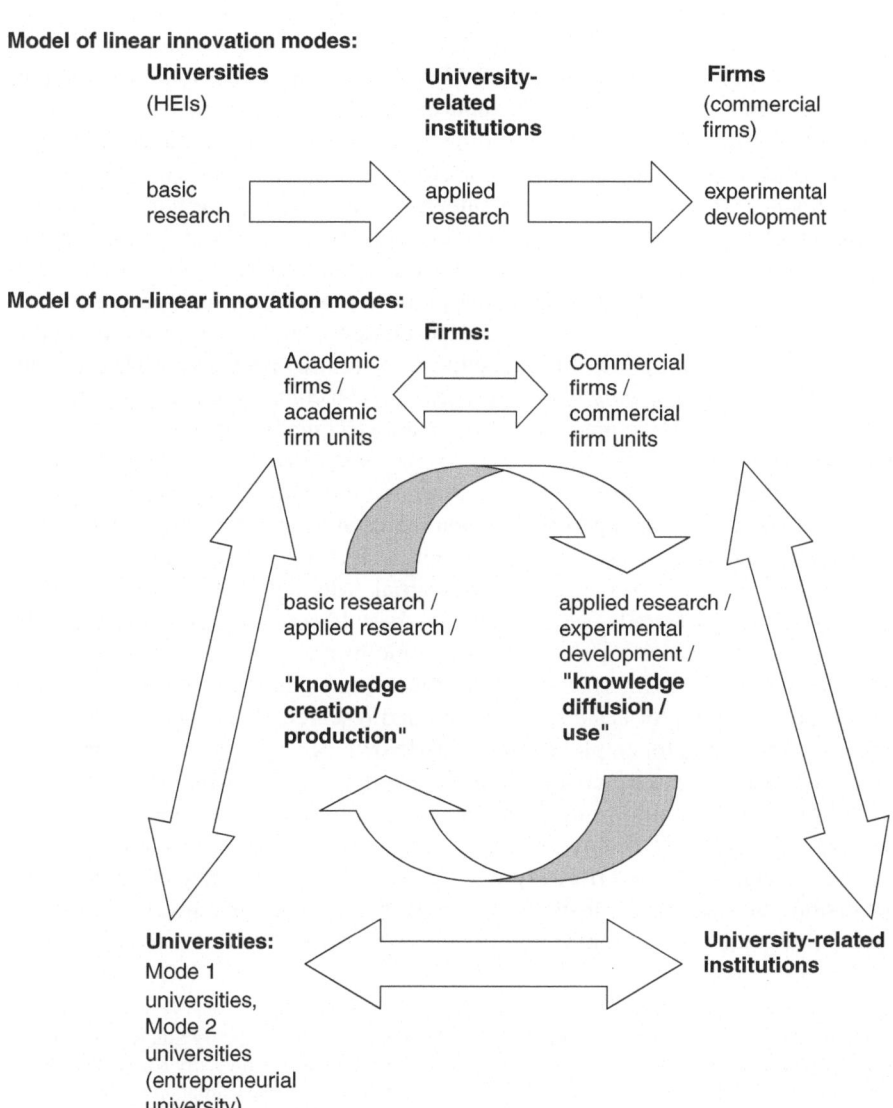

Fig. 8 Linear and nonlinear innovation modes linking together universities (Mode 1, Mode 2 and/or Mode 3 universities) with commercial and academic firms (firm units). Source: Authors' own conceptualization based on Carayannis and Campbell (2009, p. 211)

sustainable knowledge (knowledge production). Are Mode 3 universities ideal-typical concepts or are they empirical concepts? [29] Do Mode 3 universities indicate examples for ambidextrous organizations? This concept and postulated new type of Mode 3 universities (higher education institutions and subunits) pose interesting (also intellectual and epistemic) challenges for evaluation, quality assurance, and quality enhancement.[30] It is evident that the internal and external governance of higher education, also relying on evaluation and quality enhancement, must be adapted (in concepts and procedures) in context of Mode 3 universities. It also will be interesting to see and to verify later (at a later point in time), if the here presented concept of the Mode 3 university will have an influence or impact on the academic (interdisciplinary) discipline of higher education studies or research on higher education ("*Hochschulforschung*") and, furthermore, can contribute to a further theory-building and theory-development ("*Theorieentwicklung*") in reference to higher education.[31]

As important, as the entrepreneurial university or the Mode 3 university, is for us the concept of the "academic firm,"[32] which represents the complementary business organization and strategy *vis-à-vis* the entrepreneurial and Mode 3 universities. The interplay of academic firms and entrepreneurial (Mode 3) universities should be regarded as crucial for advanced knowledge-based economies and societies. The following characteristics represent the academic firm (Campbell and Güttel 2005, p. 171): "support of the interfaces between the economy and the universities"; "support of the paralleling of basic research, applied research, and experimental development"; "incentives for employees to codify knowledge"; "support of collaborative research and of research networks"; and "a limited 'scientification' of business R&D." Despite continuing important functional differences between universities and firms, also some limited hybrid overlapping may occur between entrepreneurial universities and academic firms, expressed in the circumstance that entrepreneurial universities and academic firms can engage more easily in university/business research networks. In an innovation-driven economy the business R&D is being

---

[29] For a short discussion on whether or not the IFF Faculty for Interdisciplinary Studies (*iff Fakultät für Interdisziplinäre Forschung und Fortbildung*), at the Alpen-Adria-University Klagenfurt, qualifies to represent a Mode 3 type of organization in higher education, see Campbell (2009, pp. 122, 127).

[30] On evaluation, quality management and quality enhancement in higher education in more general, see also: Blimlinger et al. 2010; Campbell 2003; Jacob 2007; Teichler 2006.

[31] The tenure-track model represents a well-established standard model for academic careers in higher education, particularly for the core faculty at universities and other higher education institutions. Already earlier in our analysis we introduced the idea of "cross-employment" or "multi-employment," where a faculty member or knowledge worker would have employment relations with different organizations or institutions (within the same sector or cross-cutting alternative sectors) at the same time. Cross-employment, therefore, allows individuals to opt for "parallel careers" within and/or outside academia. *It remains to be seen, whether cross-employment has the capability to establish itself as an additional and positively-defined role model for academic careers in higher education, in parallel to the already existing role mode of tenure-track (tenure).*

[32] The "academic firm," as a notion and concept, was first developed by Campbell and Güttel (2005).

supported and excelled when it can refer to inputs from networking of universities and firms clearly supports business R&D. The academic firm also engages in "basic business research." Of course, we always must keep in mind that academic firms and universities are not identical, because academic firms represent business units, still interested in creating commercial revenues and profits.

The *Commercial Firm* concentrates on maximizing or optimizing profit, whereas the *Academic Firm* focuses on maximizing or optimizing knowledge and innovation. While the entrepreneurial (Mode 2) university represents a partial extension of business elements to the world of academia, the academic firm could serve as an example for an extension of the world of academia to the world of business. Academic firms are knowledge-oriented, interested in engaging in networks with universities (the higher education sector), encourage "academic culture and values" to motivate their employees, allow forms of academic work (such as academic-style publishing), and support continuing education and lifelong learning of and for their employees (flexible time schemes, honoring lifelong and continued learning, and continuing education with internal career promotion).

The concept of the "academic firm" may refer *to*:

1. A whole firm
2. A subunit, subdivision, or branch of a "commercial" firm[33]
3. Certain characteristics or elements of a whole (commercial) firm

Are academic firms ideal-typical or empirical concepts? Are firms, interested in integrating principles of the commercial and academic firm, examples for ambidextrous organizations? *For the future, this may have the following challenging implication: How can or should firms balance, within their "organizational boundaries," principles of the academic and of the more traditional "commercial" firm?*

The academic firm concept is an ideal model where the degree, intensity, breadth, and depth of the Mode 3 and Quadruple Innovation Helix structures and dynamics are at their optimal level. In reality, there are cases of academic spin-offs as well as industry start-ups that approximate the academic firm modus operandi with substantial divergences and differences still. For instance, an earlier example may be that of Thermo-electron which started with one academic spin-off and evolved into an entire innovation ecosystem of entities that were built around technology solutions emerging from MIT academic research and there are many other later instances of evolved cases where there is a fusion of research, education, and innovation processes, events, and mandates (including strategic knowledge serendipity and arbitrage events as well as "happy knowledge accidents") that represent instances of an academic firm. Clearly, we are not talking about nonprofits (they may be a subset of the academic firm concept in some cases) as we focus on the substance and dynamics of the organizational forms in question and not the formalism of their organizational type (for or not-for-profit).

---

[33] In many contexts, this second option appears to be more realistic, particularly when we analyze multinational companies or corporations (MNCs) that operate in global context.

The "technology life cycles" explain why there is always a dynamic momentum in the gloCal knowledge economy and society (Tassey 2001). The "saturation tendency" within every technology life cycle demands the creation and launch of new technology life cycles, leading to the market introduction of next generation technology-based products and services. In reality, always different technology life cycles with a varying degree of market maturity will operate in parallel. To a certain extent, technology life cycles are also responsible for the cyclicality (growth phases) of a modern market economy. The perhaps shortest possible way of describing the economic thinking of Joseph A. Schumpeter is to put up the following equation: entrepreneurship, leveraging the opportunities of new technology life cycles, creates economic growth. Addressing the cyclicality of capitalist economic life, Schumpeter (1942) used the notion of the "Creative Destruction." "Mode 3" may open up a route for overcoming or transforming the destructiveness of the "creative destruction" (Carayannis et al. 2007).

2 The Conceptual Understanding of Knowledge and Innovation

Knowledge does matter: but the question is when, how, and why? Moreover, with the advancement of economies and societies, *knowledge matters even more* and in ways that are not always predictable or even controllable (e.g., see the concepts of *strategic knowledge serendipity* and *strategic knowledge arbitrage* in Carayannis et al. 2003). The successful performance of the developed *and* the developing economies, societies, and democracies increasingly depends on knowledge. One branch of knowledge develops along R&D (research and experimental development), S&T (science and technology), and innovation.[34]

2.1 Innovation Placed in Context

Discovery consists of looking at the same thing as everyone else and thinking something different

Albert Szent-Gyorgyi—Nobel Prize Winner

Innovation is a word derived from the Latin, meaning to introduce something new to the existing realm and order of things or to change the yield of resources as stated by J.B. Say quoted in Drucker (1985).

In addition, innovation is often linked with creating a sustainable market around the introduction of new and superior product or process. Specifically, in the literature

---

[34] Another branch of knowledge can be based on education and its diversified manifestations.

on the management of technology, technological innovation is characterized as the introduction of a new technology-based product into the market:

> "*Technological innovation* is defined here as a situationally new development through which people extend their control over the environment. Essentially, technology is a tool of some kind that allows an individual to do something new. A technological innovation is basically information organized in a new way. So technology transfer amounts to the communication of information, usually from one organization to another." (Tornatzky and Fleischer 1990)
>
> The broader interpretation of the term "innovation" refers to an innovation as an "idea, practice or material artifact" (Rogers and Shoemaker 1971, p. 19) adopted by a person or organization, where that artifact is "perceived to be new by the relevant unit of adoption" (Zaltman et al. 1973). Therefore, innovation tends to change perceptions and relationships at the organizational level, but its impact is not limited there. Innovation in its broader socio-technical, economic, and political context can also substantially impact, shape, and evolve ways and means people live their lives, businesses form, compete, succeed, and fail, and nations prosper or decline.

From a business perspective, an innovation is perceived as the happy ending of the commercialization journey of an invention, when that journey is indeed successful and leads to the creation of a sustainable and flourishing market niche or new market. Therefore, a technical discovery or invention (the creation of something new) is not significant to a company unless that new technology can be utilized to add value to the company, through increased revenues, reduced cost, and similar improvements in financial results. This has two important consequences for the analysis of any innovation in the context of a business organization.

First, an innovation must be integrated into the operations and strategy of the organization, so that it has a distinct impact on how the organization creates value or on the type of value the organization provides in the market.

Second, an innovation is a social process, since it is only through the intervention and management of people that an organization can realize the benefits of an innovation.

The discussion of innovation clearly leads to the development of a model, to understand the evolving nature of innovation. Innovation management is concerned with the activities of the firm undertaken to yield solutions to problems of product, process, and administration. Innovation involves uncertainty and dis-equilibrium. Nelson and Winter (1982) propose that almost any change, even trivial, represents innovation. They also suggest, given the uncertainty, that innovation results in the generation of new technologies and changes in relative weighting of existing technologies (ibid). This results in the *disruptive process* of dis-equilibrium. As an innovation is adopted and diffused, existing technologies may become less useful (reduction in weight factors) or even useless (weighing equivalent to "0") and abandoned altogether. The adoption phase is where uncertainty is introduced. New technologies are not adopted automatically but rather, markets influence the adoption rate (Carayannis 1997, 1998). Innovative technologies must propose to solve a market need such as reduced costs or increased utility or increased productivity. The markets, however, are social constructs and subject to non-innovation related criteria. For example, an invention may be promising, offering a substantial reduction on the cost of a product which normally would influence the market to accept the given innovation, but due to issues like information asymmetry (the lack of knowledge in the market concerning the invention's properties), the invention may not be readily accepted by the markets. Thus the innovation may remain an invention. If, however, the innovation is market accepted, the results will bring about change to the existing technologies being replaced, leading to a change in the relative weighting of the existing technology. This is in effect *dis-equilibrium.*

Given the uncertainty and change inherent in the innovation process, management must develop skills and understanding of the process a method for managing the disruption. The problems of managing the resulting disruption are strategic in nature. The problems may be classified into three groups: *engineering, entrepreneurial, and administrative* (Drejer 2002). This grouping correlates to the related types of innovation namely, *product, process, and administrative innovation*:

- *The engineering problem is one of selecting the appropriate technologies for proper operational performance.*
- *The entrepreneurial problem refers to defining the product/service domain and target markets.*
- *Administrative problems are concerned with reducing the uncertainty and risk during the previous phases.*

In much of the foregoing discussion, a recurring theme about innovation is that of *uncertainty*, leading to the conclusion that an effective model of innovation must include a multidimensional approach (uncertainty is defined as unknown unknowns whereas risk is defined as known knowns). One model posited as an aide to understanding is the Multidimensional Model of Innovation (MMI) (Cooper 1998). This model attempts to define the understanding of innovation by establishing three-dimensional boundaries. The planes are defined as product-process, incremental-radical, and administrative-technical. The product-process boundary concerns itself with the end product and its relationship to the methods employed by firms to produce and distribute the product. Incremental-radical defines the degree of relative strategic change that accompanies the diffusion of an innovation. This is a measure of the disturbance or disequilibrium in the market. Technological-administrative boundaries refer to the relationship of innovation change to the firm's operational core. The use of technological refers to the influences on basic firm output while the administrative boundary would include innovations affecting associated factors of policy, resources, and social aspects of the firm.

2.2   The Relationship Between Knowledge and Innovation

What is the relationship between knowledge and innovation? From our viewpoint it makes sense, not to treat knowledge and innovation as interchangeable concepts. Ramifications of this are (see Fig. 9):

Knowledge

	yes	no
yes	knowledge-based innovation or knowledge, which through innovation, is linked with society, economy and politics. Examples: Mode 1 and technology cycles in the long run, Mode 2, Triple Helix.	Innovation, taking place with no (almost no) references to knowledge. (at least with no references to research, R&D, or a research-based knowledge). Examples: management innovations in businesses, which are not R&D or technology-based.
no	Knowledge, without major references to innovation (and use). Examples: "pure research", perhaps some components of Mode 1 and of early phases of technology life cycles.	*? (Not of primary concern for our conceptual mapping here, perhaps in epistemological and philosophical terms of interest.)*

Innovation

Fig. 9 A fourfold typology about possible cross-references and interactions between "knowledge" and "innovation." Source: Authors' own conceptualization based on Carayannis and Campbell (2009, p. 213)

1. There are aspects, areas of knowledge, which can be analyzed, without considering innovation (e.g., "pure basic research" in a linear understanding of innovation).
2. Consequently, there are also areas or aspects of innovation, which are not (necessarily) tied to knowledge or a research-based knowledge. For example, see the different contributions to Shavinina (2003).
3. However, there are also areas, where knowledge and innovation coexist. These we would like to call *knowledge-based innovation*, indicating areas, where knowledge and innovation express a mutual interaction.

In the case of knowledge-referring innovation, we then can speak of innovation that deals with knowledge. Our impression is that in many contexts, when the focus falls on innovation, almost automatically this type of "knowledge-referring" or "knowledge-based" innovation is implied. Even though we will focus on this knowledge-based innovation, it still is important to acknowledge the possibilities of a knowledge without innovation, *and* of an innovation, independently of knowledge or a research-based knowledge. To further illustrate our point, the notion of the "national innovation system" (NIS, also NIS indicators) or "national system of innovation" conventionally expresses linkages to knowledge. The national innovation system as an idea and concept is being closely associated with the two scholars Bengt-Åke (1992) and Richard R. Nelson (1993).[35]

2.3 The "Mode 3" Knowledge Production System Multilevel Approach to Knowledge and Innovation: The "Multilevel Innovation Systems"

In research about the European Union (EU), references to a "multilevel architecture" are quite common (see, e.g., Hooghe and Marks 2001). Originating from this research about the EU, this "multilevel" approach is being applied in a diversity of fields, since it supports the understanding of complex processes in a globalizing world. Inspired by this, we suggest using the concept of *multilevel systems of knowledge* (see Fig. 10; see, furthermore, Carayannis and Campbell 2006a). One obvious axis, therefore, is the spatial (geographic, spatial-political) axis that expresses different levels of spatial aggregations. The national level, coinciding with the nation state (the currently dominant manifestation of arranging and organizing political and societal affairs), represents one type of spatial aggregation. Subnational aggregations fall below the nation state level, and point toward local political entities. Transnational aggregations, for example, can refer to the supranational integration process of the EU. This raises the interesting question, whether we should be prepared to expect that in the twenty-first century we will witness a proliferation of supranational (transnational) integration processes also in other world regions, possibly implying a new stage in the evolution of politics, where (small and medium-sized) nation state structures become absorbed by supranational (transnational) clusters (Campbell 1994). The highest level of transnational aggregation, we currently know, is globalization. Interestingly, the aggregation level of the term "region(s)" has never been convincingly standardized. In the context and political language of the EU, regions are understood subnationally. American scholars, on the other hand, often refer to regions in a state-transcending understanding (i.e., a region consists more than one nation states). The new term gloCal (global/local; Carayannis and Von Zedtwitz 2005) underscores the potentials and benefits of a mutual and parallel interconnectedness between different levels.

---

[35] As an example for a reviewing of innovation policy in context of a national innovation system, see the analysis of Guy Ben-Ari (2006) on Israel.

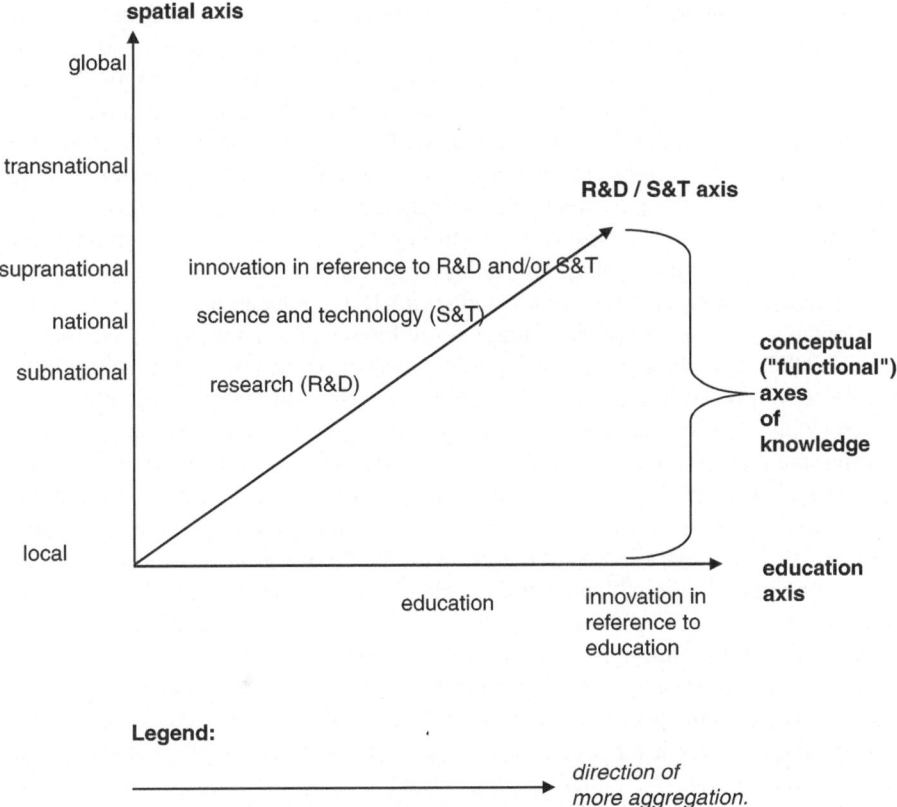

Fig. 10 A "three-dimensional" modeling of knowledge (and innovation) in a multilevel system understanding: axis of spatial aggregation, axis of R&D aggregation, and axis of education aggregation (the "multilevel innovation systems"). Source: Authors' own conceptualization, adapted from Carayannis and Campbell (2009, p. 215)

Despite the importance of this spatial axis, we wish not to exhaust the concept of multilevel systems of knowledge and innovation with spatial-geographic metaphors. *For us, the concept of multilevel innovation systems*[36] *is clearly more than a primarily "spatial" or geographic concept. We suggest adding on nonspatial axes of aggregation. These we may call conceptual (functional) axes of knowledge and innovation.* In that context, two axes certainly are pivotal: education and research (R&D, research and experimental development). For research, the level of aggregation can

---

[36] A possible acronym here may be: MLIS or MLISs.

develop accordingly: R&D; S&T (science and technology)[37]; and R&D-referring innovation, involving a whole broad spectrum of considerations and aspects. Obviously, every "axis direction" of further aggregation—as demonstrated here for R&D—depends on a specific conceptual understanding. Should, for example, a different conceptual approach for defining S&T be favored, then the sequence of aggregation might change. (Concerning the education axis, for the moment, we want to leave it to the judgment of other scholars, what here meaningful terms at different levels of aggregation may be.) In Fig. 10, we present a three-dimensional visualization of a multilevel system of knowledge, combining one spatial with two nonspatial (conceptual) axes of knowledge (R&D and education).

How many nonspatial (conceptual) axes of knowledge can there be? We focused on the R&D and education axes. By this, however, we do not want to imply that there may not be more than two conceptual axes. Here, at least in principle, a multitude or diversity of conceptual model-building approaches are possible and also appropriate. Perhaps, we even could integrate "innovation" as an additional conceptual axis, following the aggregation line from local to national and transnational innovation systems. We then would have to contemplate what the relationship is between such an "extra innovation axis" with the "innovation" of the research and education axes. "Regional" innovation could cross-reference local and transnational innovation systems, implying even gloCal innovation systems and processes that simultaneously link through different aggregation levels.

We already discussed the conceptual boundary problems between knowledge and innovation. One approach, how to balance ambiguities in this context, is to acknowledge that a partial conceptual overlap exists between a *knowledge-centered* and *innovation-centered* understanding. Depending on the focus of the preferred analytical view, the same "element(s)" can be conceptualized as being part of a knowledge or of an innovation system. Concerning knowledge, we pointed to some of the characteristics of multilevel systems of knowledge, underscoring the understanding of aggregation of spatial and nonspatial (conceptual) axes. Introducing multilevel systems of knowledge also justifies speaking of multilevel systems of innovation, developing the original concept of the national innovation system (Lundvall 1992; Nelson 1993) further. For example, the spatial axis of aggregation of knowledge (Fig. 10) also applies to innovation. Of course, also Lundvall (1992, pp. 1, 3) explicitly stresses that national innovation systems are permanently challenged (and extended) by regional as well as global innovation systems. But, paraphrasing Kuhlmann (2001, pp. 960–961), as long as nation state-based political systems exist, it makes sense to acknowledge national innovation systems. In a spatial (or geographic) understanding, the term multilevel systems of innovation already is being used (Kaiser and Prange 2004, pp. 395, 405–406; Kuhlmann 2001, pp. 970–971, 973). However, only more recently has it been suggested to extend this multilevel aggregation approach of innovation also to the nonspatial axes of innovation

---

[37] In that context also the mutual overlapping between R&D, S&T and ICT (information and communication technology) should be stressed.

(Campbell 2006a; Carayannis and Campbell 2006a). Therefore, multilevel systems of knowledge as well as multilevel systems of innovation are based on spatial and nonspatial axes. A further advantage of this multilevel systems architecture is that it results in a more accurate and closer-to-reality description of processes of globalization and gloCalization. For example, internationalization of R&D cross-cuts these different multilevel layers and links together organizational units of business, academic, and political actors at national, transnational, and subnational levels (Von Zedtwitz and Heimann 2006). One interpretation of R&D internationalization emphasizes how different subnational regions and clusters cooperate on a global scale, creating even larger transnational knowledge clusters.

The concept of the "sectoral systems of innovation" (SSI) cross-cuts the logic of the multilevel systems of innovation or knowledge.[38] A sector often is being understood in terms of the industrial sectors. Sectors can perform locally/regionally, nationally, and transnationally. Reviews of SSIs often place a particular consideration on: knowledge and technologies; actors and networks; furthermore institutions. Malerba (2004a, p. i) recommends that analyses of sectoral systems of innovation should include "the factors affecting innovation, the relationship between innovation and industry dynamics, the changing boundaries and the transformation of sectors, and the determinants of the innovation performance of firms and countries in different sectors."

2.4   Linear Versus (and/or) Nonlinear Innovation Models (Modes)

Is the *linear model of innovation* still valid? In an ideal typical understanding the linear model states: first there is basic research, carried out in a university context. Later on, this basic research is converted into applied research, and moves from the university to the university-related sectors. Finally, applied research is translated into experimental development, carried out by business (the economy). What results is a *first–then relationship*, with the universities and/or basic research being responsible for generating the new waves of knowledge creation, which are, later on, taken over by business, and where business carries the final responsibility for the commercialization and marketing of R&D. National (multilevel) innovation systems, operating primarily on the premises of this linear innovation model, obviously would be disadvantaged: the time horizons for a whole R&D cycle, to reach the markets, could be quite extensive (with negative consequences for an economy, operating in the context of rapidly intensifying global competition). Furthermore, the linear innovation model exhibits serious weaknesses in communicating user preferences from the market end back to the production of basic research. In addition, how should the tacit knowledge of the users and markets be reconnected back

---

[38] On *sectoral systems of innovation*, see also in greater detail in later sections of our analysis.

to basic research? In the past, after 1945, the U.S.A. was regarded as a prototype for the linear innovation model system, with a strong university base, from where basic research gradually would diffuse to the sectors of a strong private economy, without the intervention of major public innovation policy programs (see Bush 1945, Chapter "The Importance of Basic Research"). As long as the U.S.A represented the world-leading national economy, this understanding was sufficient. But with the intensification of global competition, also the demand for shortening the time horizons from basic research to the market implementation of R&D increased (OECD 1998, pp. 179–181, 185–186). In the 1980s, Japan in particularly heavily pressured the U.S.A. In the 2000s, global competition within the triad of the U.S.A., Japan, and the EU escalated further, with China and India emerging as new competitors in the global context. In a nutshell, further-going economic competition and intrinsic knowledge demands challenged the linear innovation model.

As a consequence, we can observe a significant proliferation of *nonlinear innovation models*. There are several approaches to nonlinear innovation models. The "chain-linked model," developed by Kline and Rosenberg (1986; cited according to Miyata 2003, p. 716; see furthermore Carayannis and Alexander 2006), emphasizes the importance of feedback between the different R&D stages. Particularly, the coupling of marketing, sales, and distribution with research claims to be important. "Mode 2" (Gibbons et al. 1994, pp. 3–8, 167) underscores the linkage of production and use of knowledge, by referring to the following five principles: "knowledge produced in the context of application"; "transdisciplinarity"; "heterogeneity and organizational diversity"; "social accountability and reflexivity"; and "quality control" (furthermore, see Nowotny et al. 2001, 2003).[39] Metaphorically speaking, the *first–then* sequence of relationships of different stages within the linear model becomes replaced by a *paralleling* of different R&D activities (Campbell 1995, p. 31, 2000, p. 139–141). Paralleling means: (1) linking together in real time different stages of R&D, for example basic research and experimental development, and/or (2) linking different sectors, such as universities and firms. Is this new "paralleling" in R&D also being supported epistemically (epistemologically) by a paralleling of *if–then and if–if relations in causality* (causal thinking) (Campbell 2009, p. 123)? The "Triple Helix" model of Etzkowitz and Leydesdorff (2000, pp. 109, 111) stresses the interaction between academia, state, and industry, focusing consequently on "university–industry–government relations" and "trilateral networks and hybrid organizations." Carayannis and Laget (2004, p. 17, 19) emphasize the importance of cross-national and cross-sectoral research collaboration, by testing these propositions for transatlantic public–private R&D partnerships. Anbari and Umpleby (2006, pp. 27–29) claim that one rationale, for establishing research networks, lies in the interest of bringing together knowledge producers, but also practitioners, with "complementary skills." Etzkowitz (2003) speaks also of the "entrepreneurial

---

[39] Should we add a further comment to the concepts of Mode 1 and Mode 2, it would be interesting to consider, how Mode 1 and Mode 2 relate to the notions of "Science One" and "Science Two," which were developed by Umpleby (2002).

university." An effective coupling of university research and business R&D demands, furthermore, the complementary establishment of the entrepreneurial university and the "academic firm" (Campbell and Güttel 2005, pp. 170–172). Extended ramifications of these discourses also refer to the challenge of designing proper governance regimes for the funding and evaluation of university research (Geuna and Martin 2003; see, furthermore, Shapira and Kuhlmann 2003, and Campbell 1999, 2003). Furthermore, this imposes consequences on structures and performance of universities (Pfeffer 2006). Interesting is also the concept of "democratizing innovation." With this concept, Eric von Hippel proposes a "user-centric innovation" model, in which "lead users" represent "innovating users," who again contribute crucially to the performance of innovation systems. "Lead users" can be individuals or firms. Users often innovate, because they cannot find on the market, what they want or need (Von Hippel 2005; also, Von Hippel 1995). Nonproprietor knowledge, such as the "open source" movement in the software industry (Steinmueller 2004, p. 240), may be seen as successful examples for gloCally self-organizing "user communities."

Put in summary, one could set up the following hypothesis for discussion: while Mode 1 and perhaps also the concept of "Technology Life Cycles"[40] appear to be closer associated with the linear innovation model, the Mode 2 and Triple Helix knowledge modes have more in common with a nonlinear understanding of knowledge and innovation. At the same time we should add that national (multilevel) innovation systems are challenged by the circumstance that several technology life cycles, at different stages of market maturity (closeness to commercial market introduction), perform in parallel. This parallel as well as sequentially time-lagged unfolding of technology life cycles also expresses characteristics of Mode 2 and of nonlinear innovation, because organizations (firms and universities) often must develop strategies of simultaneously cross-linking different technology life cycles. Universities and firms (commercial and academic firms) must balance the nontriviality of a fluid pluralism of technology life cycles.

2.5 Extending the "Triple Helix" to a "Quadruple Helix" Model of Knowledge and Innovation

In their own words, Etzkowitz and Leydesdorff (2000, p. 118) say that the "Triple Helix overlay provides a model at the level of social structure for the explanation of Mode 2 as an historically emerging structure for the production of scientific knowledge, and its relation to Mode 1." Triple Helix is very powerful in describing and explaining the helices dynamics of "university–industry–government relations" that drives knowledge and innovation in the gloCal knowledge economy and society. We suggest that advanced knowledge-based economy and advanced democracy have

---

[40] Concerning a further-going discussion of the Technology Life Cycles, see: Cardullo 1999; Tassey 2001.

increasingly similar features, in the sense of combining and integrating different knowledge modes and different political modes.[41] Modern political science claims that democracy and politics develop along the premises of a "media-based democracy." Fritz Plasser (2004, pp. 22–23) offers the following description for media-based democracy: media reality overlaps with political and social reality; perception of politics primarily through the media; and the laws of the media system determining political actions and strategies. Politics may convert from a "parliamentary representative" to a "media presenting" democracy, where "decision" politics moves to a "presentation" politics. Ramifications of the "multimedia information society" clearly impact "political communication" (see also Plasser and Plasser 2002).

The "fourth helix" of the Quadruple Helix refers to this "media-based and culture-based public" as well as to "civil society" (see again Fig. 2). Knowledge and innovation policies and strategies must acknowledge the important role of the "public" for a successful achieving of goals and objectives. On the one hand, public reality is being constructed and communicated by the media and media system. On the other hand, the public is also influenced by culture and values. Knowledge and innovation policy should be inclined to reflect the dynamics of "media-based democracy," to draft policy strategies. Particularly when we assume that traditional economic policy gradually (partially) converts into innovation policy, leveraging knowledge for economic performance and thus linking the political system with the economy, then innovation policy should communicate its objectives and rationales, via the media, to the public, to seek legitimation (legitimacy) and justification (see Fig. 11; furthermore, see Carayannis and Campbell 2006a, p. 18, 2006b, p. 335). Also the PR (public relation) strategies of companies, engaged in R&D, must reflect on the fact of a "reality construction" by the media. Culture and values also express a key role. Cultural artefacts, such as movies, can create an impact on the opinion of the public and their willingness, to support public R&D investment. Some of the technical and engineering curricula at universities are not gender-symmetric, because a majority of the students are male. Trying to make women more interested in enrolling in technical and engineering studies would imply also changing the "social images" of technology in society. The sustainable backing and reinforcing of knowledge and innovation in the gloCal knowledge economy and society requires a substantive supporting of the development and evolution of "innovation cultures" (Kuhlmann 2001, p. 954). *Therefore, the successful engineering of knowledge and innovation policies and/or strategies leverages the self-logic of the media system and leverages or alters culture and values.* Etzkowitz and Leydesdorff, in their stated quote, emphasize their intention that the Triple Helix model should help displaying patterns of "social structure." This in fact provides a rationale why a fourth

---

[41] A political mode could be seen as a particular political approach (clustering political parties, politicians, ideologies, values, and policies) to society, democracy, and the economy. Conservative politics, liberal politics or social democratic politics could be captured by the notion of a "political mode".

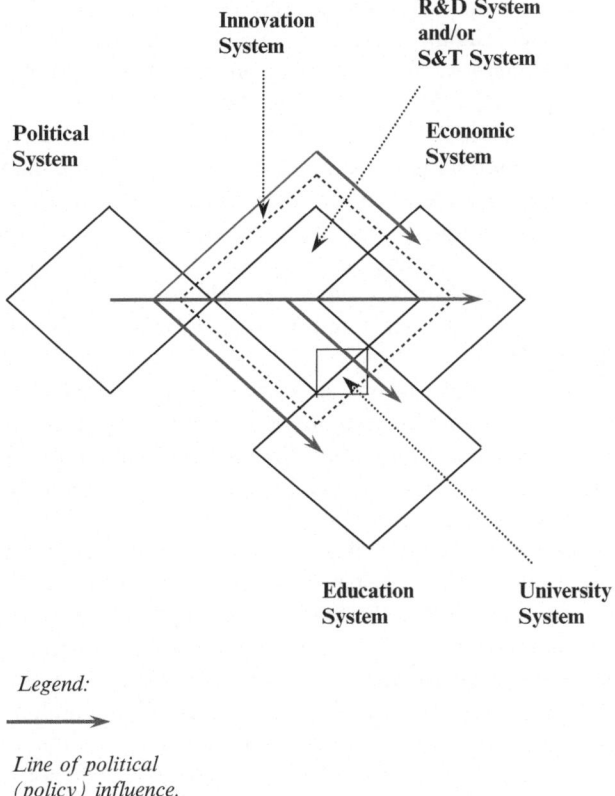

Fig. 11 Different societal systems: lines of political (policy) influence. Source: Carayannis and Campbell (2006a, p. 18, Figs. 1–7)

helix of "media-based and culture-based public" could serve as a useful analytical tool, providing additional insights.

2.6   Coexistence and Co-evolution of Different Knowledge and Innovation Paradigms

Discussing the evolution of scientific theories, Thomas S. Kuhn (1962) introduced the concept of *paradigms*. Paradigms can be understood as basic fundamentals, upon which a theory rests. In that sense paradigms are axiomatic premises, which guide a theory; however, they cannot be explained by the theory itself: but paradigms add to the explanatory power of theories that are interested in explaining the (outside) world. Paradigms represent something like beliefs. According to Kuhn, there operates an evolution of scientific theories, following a specific pattern: there are periods of "normal science," interrupted by intervals of "revolutionary science," again converting over into "normal science," again challenged by "revolutionary science," and so on (Carayannis 1993, 1994, 2000, 2001; see also Umpleby 2005, pp. 287–288). According to Kuhn, every scientific theory, with its associated paradigm(s), has only a limited capacity for explaining the world. Confronted with phenomena, which cannot be explained, a gradual modification of the same theory might be sufficient. However, at one point a revolutionary transformation is necessary, demanding that a whole set of theories/paradigms will be replaced by new theories/paradigms. For a while, the new theories/paradigms are adequately advanced. However, in the long run, these cycles of periods of normal science and intervals of revolutionary science represent the dominant pattern.

Kuhn emphasizes this shift of one set of theories and paradigms to a new set, meaning that new theories and paradigms represent not so much an evolutionary offspring, but actually replace the earlier theories and paradigms. While this certainly often is true, particularly in the natural sciences, we want to stress that there also can be a *coexistence and co-evolution of paradigms* (and theories), implying that paradigms and theories can mutually learn from each other. Particularly in the social sciences this notion of coexistence and co-evolution of paradigms might be sometimes more appropriate than the replacement of paradigms. For the social sciences, and politics in more general, we can point toward the pattern of a permanent mutual contest between ideas. Stuart A. Umpleby (1997, p. 635), for instance, emphasizes the following aspect of the social sciences very accurately: "Theories of social systems, when acted upon, change social systems." Not only (social) scientific theories refer to paradigms, also other social contexts or factors can be understood as being based on paradigms: we can speak of ideological paradigms, or of policy paradigms (Hall 1993). Another example would be the long-term competition and fluctuation between the welfare-state and the free-market paradigms (with regard to the metrics of left-right placement of political parties in Europe, see Volkens and Klingemann 2002, p. 158).

These different modes of innovation and knowledge creation, diffusion, and use, which we discussed earlier, certainly qualify to be understood also as linking to *knowledge paradigms*. Because knowledge and innovation systems clearly relate to the context of a (multilevel) society, the (epistemic) knowledge paradigms can be regarded as belonging to the "family of social sciences." Interestingly, Mode 2 addresses "social accountability and reflexivity" as one of its key characteristics (Gibbons et al. 1994, pp. 7, 167–168). In addition to the possibility that a specific knowledge paradigm is replaced by a new knowledge paradigm, the relationship between different knowledge and innovation modes may often be described as an ongoing and continuous interaction of a dynamic co-existence and (over time) a co-evolution of different knowledge paradigms. This reinforces the understanding that, in the advanced knowledge-based societies and economies, linear and nonlinear innovation models can operate in parallel.

2.7   The "Co-opetitive" Networking of Knowledge Creation, Diffusion, and Use

Knowledge systems are highly complex dynamic and adaptive. To begin with, there exists a conceptual (hybrid) overlapping between multilevel knowledge and multilevel innovation systems. Multilevel systems process simultaneously at the global, transnational, national, and subnational levels, creating gloCal (global and local) challenges. Advanced knowledge systems should demonstrate the flexibility of integrating different knowledge modes; on the one hand, combining linear and nonlinear innovation modes; on the other hand, conceptually integrating the modes of Mode 1, Mode 2, and Triple Helix (for an overview of Mode 1, Mode 2, Triple Helix, and Technology Life Cycles, see Campbell 2006a, pp. 71–75). This displays the practical usefulness of an understanding of a co-existence and co-evolution of different knowledge paradigms, and what the qualities of an "innovation ecosystem" could or even should be. The elastic integration of different modes of knowledge creation, diffusion, and use should generate synergistic surplus effects of additionality. Hence for advanced knowledge systems, networks and networking are important (Carayannis and Alexander 1999b; Carayannis and Campbell 2006b, pp. 334–339; for a general discussion of networks and complexity, see also Rycroft and Kash 1999).

How do networks relate to *cooperation and competition?* "Co-opetition," as a concept (Brandenburger and Nalebuff 1997), underscores that there can always exist a complex balance of cooperation and/or competition. Market concepts emphasize a competitive dynamics process between (1) forces of supply and demand, and the need of integrating (2) market-based as well as resource-based views of business activity. To be exact, networks do not replace market dynamics; thus they do not represent an alternative to the market-economy-principle of competition. Instead, networks apply a "co-opetitive" rationale, meaning: internally, networks are based primarily on cooperation, but may also allow a "within" competition. The relationship between different networks can be guided by a motivation for cooperation. However, in practical terms, *competition in knowledge and innovation often will be carried out between different and flexibly configured networks. While a network cooperates internally, it may compete externally.* In short, "co-opetition" should be regarded as a driver for networks, implying that the specific content of cooperation and competition is always decided in a case-specific context.

3   Sectoral Systems of Innovation and Technology Dynamics

Already earlier we introduced and referred to the concept of the *Sectoral Systems of Innovation* (Malerba 2004a) that cross-cut and complement the *architecture of the multilevel systems of innovation* (see again the Figs. 2 and 10). In the following section of this chapter, we focus more specifically on sectoral systems of innovation and on the technology dynamics in four specific technology fields (industrial sectors): the software sector; the pharmaceutical sector; the chemical sector; and the machine tool sector. For that purpose, we review and summarize concretely studies that focused on capturing the sectoral innovation momentum. Sectoral systems of innovation represent a well-established field of study, for which Franco Malerba (2004a) collected conclusive empirical evidence. *Sectoral systems of innovation, furthermore, can be regarded as an approach that displays, emphasizes, and makes transparent specific behavioral patterns in different technology fields, thus acknowledging the increasing heterogeneity and diversity (pluralism) of knowledge and innovation in advanced knowledge economy and knowledge society.*

3.1   Sectoral Systems of Innovation and Technology Dynamics in IT/ICT (Information Technology/Information and Communication Technology) in the Software Sector

The software sector we will use as an example for describing more specifically some of the developments in IT/ICT. Two major "coordinating mechanisms" here are the "role of technological standards" and the "role of dominant competitors" in the different segments of software industry. One objective of such coordination mechanisms is to realize "system integration" at the "global level." The U.S.A. occupies, also in global terms, a very strong position in the software sector, coined in the phrase of a "US dominance of the very important, but limited, packaged software market for generic application—the global software product" (Steinmueller 2004, p. 240). To a certain extent, the leading U.S. firms are here in a "winner takes all" position, based on the circumstances that software markets are often "global in scope" and the "costs of 'localizing' products in language and culture" again are quite often low. European (and other non-U.S.) firms depend frequently on the brokering powers of the dominant competitors to (1) de facto establish their *own company-based technology standards* as the *world technology standards* (as an example, think here of Microsoft with "Windows" and subsequent office packages) or to (2) coordinate the establishment of standards, which again are in the favor of the dominant competitors. Alternative options may be "open standards" and their support, also indicating routes for strategic decision-making of European, non-European, but even some American companies. The whole spectrum of software development or software solutions may be classified into "packaged software"/"global package software" (with the example of several Microsoft software products), "middleware," "nongeneric software applications"/"situated and embedded software" (quasi stand-alone), and "open source" software. These nongeneric software applications focus on specific problem solutions or applications, but are challenged by features such as "software marketing" and "distribution." The Internet certainly introduced here opportunities for small and medium-sized software firms. Furthermore, and generally speaking, the separation lines between "operating systems" and "application software" become increasingly blurred (Steinmueller 2004, pp. 238–241; Malerba 2004b, pp. 469–470). "Middleware" and the "open source" movement represent two interesting forms of sectoral innovation; therefore they should be described here in more detail:

1. *Middleware*: Middleware can be defined as an "integrated software solution" (ISS).[42] On the one hand, middleware is "generic," because it is interested in and also capable of reaching larger user communities. On the other hand, middleware is also "situated," since its application requires considerable efforts of "user specification" and "customization," often involving the consultation of professional services. Examples for the suppliers of middleware are European companies such as SAP and Software AG, and the American companies Oracle and Microsoft. Interestingly, the creation of such middleware reflects a sophisticated division of labor. The lead company develops and sponsors a "platform" or "architecture," which represents proprietary knowledge. In principle, such platform architectures could also be designed by a consortium of companies. Then these platforms are filled and complemented with "modules," developed by "specialized software firms" that reflect "industry-specific requirements" of "users." What results is a complex network of interwoven firm activities. "Leading firms in this market are engaged in assembling networks of supplier firms that are willing to provide modules that operate within the architecture sponsored by a specific individual leading firm" (Steinmueller 2004, p. 216). The deliberate *opening* of such software-based platforms to special software-module suppliers should also recognize the trend that application demands of different user communities become more and more heterogeneous, so it would be increasingly difficult for one (even leading) software firm to still address all customer needs. This *platform approach* is not only being endorsed by the software industry alone, but also in the hardware industry we can observe the formation of hardware platforms (Malerba 2004b, pp. 475–476).

2. *Open source*: New "distribution channels" and the global diffusion of the Internet lay the foundation for the successful impact and spreading of the "open source" movement for software creation and software production. Steinmueller (2004, p. 240) even goes so far, putting forward the assertion to qualify open source as a "global development originating in Europe." Open source represents for many firms an exit option of circumventing the market powers of some dominant software companies with regard to proprietary software. Open source, therefore, is principally of interest to all firms and user communities that do not occupy a leading market position. This represents one realistic strategy option for the European software sector, which produced not as many competitive and internationally domineering software companies as is the case in the U.S.A. Open source also offers particular opportunities and potential benefits to the emerging national innovation systems of the newly industrializing countries (NICs). Challenges for open source, of course, are as follows: (a) the development of

---

[42] For a short review on "embedded systems research" in Austria, see the analysis by Prem (2005).

"effective user interfaces" for mass markets; and (b) the development of "business models,"[43] not questioning the generally free access of users to open source software. Steinmueller (2004, p. 240) underscores two important advantages of open source: (a) Open source can promote "open standards for information representation," with the one possible consequence that competition then may focus more specifically on designing "tools for information creation, analysis, and communication." (b) Programmers (young programmers) can develop their own skills, when referring to open source software by developing software-rooted problem solutions for comprehensive commercial applications. This, in fact, could also be seen as a contribution to a further enhancement of the human resource base of the European software industry. Open source innovation in the software sector quite obviously cross-links with features such as "democratizing innovation," as is being indicated by Eric von Hippel (2005). Open source-based software, in context of a further spreading and diffusion of Internet, qualifies as a prime successful example for a "user-centered" "democratization of innovation" (Von Hippel 2005, pp. 2, 177).

3.2   Sectoral Systems of Innovation and Technology Dynamics in Life Sciences, Biotechnology, and the Pharmaceutical Industry

In the following two subsections, we summarize some of the key findings about sectoral innovation in the two industrial sectors of pharmaceuticals and chemistry.

3.2.1   Sectoral Innovation and Technology Dynamics in the Pharmaceutical Sector

The pharmaceutical sector is being driven by phenomena such as "hybridization" of "organizational forms," where, partially, large corporations, NBFs (new biotechnology firms), and universities display similar behavioral patterns. Here, processes of a "division of labor" and of a "vertical and horizontal integration" take place simultaneously at the same time. Agents (actors) permanently redefine their positions, roles, and functions in complex networks, constantly changing the "space." "Thus, the pharmaceutical industry example demonstrates both chaotic behavior in the system as well as quite positive outcomes for firms and for innovative activities, at least during certain periods" (McKelvey et al. 2004, p. 113). Crucial for network-style interactions in the pharmaceutical sector are as follows: "university-industry interaction,"

---

[43] Here again aspects and criteria such as the advertisement or "attention economy" (Davenport and Beck 2001) could be mentioned.

the competitiveness of "basic science," the availability and mobility of "pooled and skilled labor forces," access opportunities to "venture capital," and, quite obviously, "regulation" regimes and "competition" patterns. The increasing complexity of the "search space" is being paraphrased by using and referring to the metaphor of an "explosion." "Exploration" turned into a more "difficult," "costly," but also "important" enterprise. The evolving and increasing complexity of the search space, one may postulate, also has lead to a hybridization of network-based interactions, because hybrid networks appear as a proper organizational pattern match for dealing with complexity. "Given the complexity of the space to be searched and the speed at which new hypotheses and techniques are generated, no individual firm can hope to be able to explore and to keep control of more than a small subset of such space" (McKelvey et al. 2004, p. 115). The U.S. innovation system has had and still has the flexibility, to deal with these challenges dynamically. "The US system was able to evolve, building on some of its typical features, into a highly decentralized but at the same time strongly integrated structure, which appears to be rather successful in combining exploration and exploitation" (McKelvey et al. 2004, p. 115). Contrarily, at least in the past, the European pharmaceutical industry was challenged by an orientation toward "domestic markets" and "fragmented research systems," implying a lack of competition and "insufficient degrees of organizational integration" (McKelvey et al. 2004, p. 116).

When interpreting knowledge patterns as a source, then, until 1945, pharmaceuticals did not differ that much from chemicals. After 1945, however, pharmaceuticals converted into a highly R&D-intensive business, with increasing complexity degrees of their "search space," leading to the already stated consequence that "nowadays no individual firm can gain control of more than a subset of the search space" (Malerba 2004b, p. 468). Innovations in the pharmaceuticals depend increasingly on a science base, scientific research potentials, and collaborations with university institutions.[44] "Innovativeness" and "competitiveness" criteria of the larger pharmaceutical firms are determined by their ability of simultaneously interacting with academic science organizations as well as with other very "specialized innovative firms": "As of now, the pharmaceutical/biotechnology sectoral system has a structure of innovative actors that include large firms, NBFs, small firms, and individuals (such as scientists or NBF entrepreneurs)" (Malerba 2004b, p. 472).

3.2.2   Sectoral Innovation and Technology Dynamics in the Chemical Sector

The chemical sector is being populated by firms with large R&D departments and can be characterized to have developed rather frequently networks with universities and other academic research organizations. The chemical industry represents also a

---

[44] This may create opportunities for the *Academic Firm*. We introduced the academic firm as a concept already earlier.

"science-based industry." While the knowledge base of chemistry expresses features of a sequential series of "discontinuities," there is also the "continuity" of some of the bigger and larger firms. "Moreover, there has been a process of *coevolution* between small and large companies, markets, research institutions, and other organizations, with firms playing the central role within the chemical sectoral system" (Cesaroni et al. 2004, p. 150). Some key characteristics are as follows: (1) knowledge and R&D determine growth potentials of chemical firms and competitive advantages to a large extent; (2) diversified networks at different levels, between firms and universities as well as between firms and firms, are crucial; (3) a specific division of labor developed, also between "chemical companies" and "technology suppliers" ("specialized engineering firms"); (4) and cross-linkages between firms and the users of chemical products clearly gained in importance.

There is this interesting tendency that larger chemical firms, more and more, source out and license out "proprietary technologies to other firms." This contributes also to a general diffusion of knowledge. One basis (prerequisite) for all of this is that progress in chemistry has also lead to an increased "codifiability of chemical knowledge" (Cesaroni et al. 2004, p. 151). Particularly the breakthroughs in "polymer chemistry" and "chemical engineering" were important for enhancing such linkages. Internal R&D processes of firms, therefore, can be complemented more easily via such external knowledge links. This, in fact, "allowed the separation of process innovation from product innovation: process innovation became a commodity that could be traced. In general, one could claim that these changes led to a transformation of firms' learning processes away from trial and error procedures to a science-based approach to industrial research" (Malerba 2004b, p. 469). The "separability" and "transferability" of knowledge developed and aligned here hand in hand. All together, the chemical sector knows and applies three types of networks, with differing frequencies in the subfields of chemistry: "inter-firm," "university-industry" and "user–producer."

3.3 Sectoral Systems of Innovation and Technology Dynamics in the Machine Tool Sector

Key elements of the machine tool industry are "strong regional sectoral linkages" and a close coupling of regional production patterns with users. The previous innovation system in machine tools could be characterized by the following features (Wengel and Shapira 2004, p. 280): "closed"; "regional and national"; "mechanically based"; "incremental"; "producers linked with users"; and "tacit knowledge."

This configuration, however, has entered a phase of re-shaping and transformation in recent years. (1) There clearly is more of a need to engage in research (R&D) collaborations, so this sector develops more research-based. (2) IT turned into an additionally important component for the machine tool industry. (3) Also in machine tools, innovation increasingly relies on a science base. (4) Faster responses to markets and customer-needs turned into a must. (5) Furthermore, aspects of internationalization gain in importance. There are expectations that within the regional clusters of machine tool industry, specific and more patterns of international cooperation activity will be integrated, at least for larger firms. (6) New technology-based firms play now a greater role in the machine tool sector than before. (7) Additionally, machine tool firms from China, Taiwan, and South Korea could change and shift sectoral innovation patterns, since these firms develop "stronger capabilities in research and innovation and are augmenting their human capital capabilities" (Wengel and Shapira 2004, p. 284). For the newly emerging and establishing innovation regimes in machine tools, Wengel and Shapira (2004, p. 280), therefore, recommend the following references and principles: "more open"; "partnerships"; "regional to international in scope"; "based on new technology"; "information-intensive"; "linkages with research centers, producers, and users"; and "increased codified knowledge." According to Malerba (2004b, pp. 470–471), the originally and primarily "incremental" knowledge base of machine tools switched in favor of an increasingly "systemic" knowledge base, where products become more "modularized and standardized." Malerba (2004b, p. 476) emphasizes, how "new actors" (e.g., communities, based on new technologies, such as nanotechnology) enter the scene, and how market mechanisms are being more frequently introduced in arrangements, such as "customer-supplier interactions."

4 Conclusion

"Until philosophers are kings, or the kings and princes of this world have the spirit and power of philosophy,... cities will never have rest from their evils—no, nor the human race as I believe..." [emphasis added]

[Plato, The Republic, Vol. 5, p. 492]

The empires of the future are the empires of the mind

Winston Churchill, 1945

This chapter that builds on prior research and publications on the Mode 3 knowledge production system and the related concepts of the Quadruple and Quintuple Innovation Helixes, reflects insights and lessons learned from ongoing research in the related areas:

1. The role, presence, and significance of higher order learning as first outlined in the doctoral thesis of the first author in 1994 (*The Strategic Management of Technological Learning*, Carayannis, July 1994 [see also Carayannis 2001])
2. The need for a more holistic and humble (open-minded) approach in modeling the world
3. The need for hybrid, quantitative as well as qualitative research tools and protocols in trying to map and model such phenomena, processes, and dynamics which we have been enacting during the last 5 years

The "Mode 3" systems approach for knowledge creation, diffusion, and use emphasizes the following key elements (Carayannis and Campbell 2006c):

1. *GloCal multilevel knowledge and innovation systems*: Because of its comprehensive flexibility and explanatory power, systems theory is regarded as suitable for framing knowledge and innovation in the context of multi-level knowledge and innovation systems (Carayannis and Von Zedtwitz 2005; Carayannis and Campbell 2006c; Carayannis and Sipp 2006). GloCal expresses the simultaneous processing of knowledge and innovation at different levels (e.g., global, national, and subnational; see, furthermore, Gerybadze and Reger 1999, and Von Zedtwitz and Gassmann 2002), and also refers to stocks and flows of knowledge with local meaning and global reach. Knowledge and innovation systems (and concepts) express a substantial degree of hybrid overlapping, meaning that often the same empirical information or case could be discussed under the premises of knowledge or innovation.

2. *Elements/clusters and rationales/networks*: In a theoretical understanding, we pointed to the possibility of linking the "elements of a system" with clusters and the "rationale of a system" with networks. Clusters and networks are common and useful terms for the analysis of knowledge.

3. *Knowledge clusters, innovation networks, and "co-opetition"*: More specifically, we emphasize the terms of "knowledge clusters" and "innovation networks" (Carayannis and Sipp 2006). Clusters, from an ultimate perspective, by taking demands of a knowledge-based society and economy seriously for a competitive and effective business performance, should be represented as knowledge configurations. Knowledge clusters, therefore, represent a further evolutionary development of geographical (spatial) and sectoral clusters. Innovation networks, internally driving and operating knowledge clusters or cross-cutting and cross-connecting different knowledge clusters, enhance the dynamics of knowledge and innovation systems. Networks always express a pattern of "co-opetition," reflecting a specific balance of cooperation and competition. Intra-network and inter-network relations are based on a mix of cooperation and competition, i.e., co-opetition (Brandenburger and Nalebuff 1997). When we speak of competition, it often will be a contest between different network configurations.

4. *Knowledge fractals*: "Knowledge fractals" emphasize the continuum-like bottom-up and top-down progress of complexity. Each subcomponent (sub-element) of a knowledge cluster and innovation network can be displayed as a micro-level sub-configuration of knowledge clusters and innovation networks (see Fig. 12). At the same time, one can also move upward. Every knowledge cluster and innovation network can also be understood as a subcomponent (sub-element) of a larger macro-level knowledge cluster or innovation network in other words, innovation meta-networks and knowledge meta-clusters (see again Fig. 12).[45]

---

[45] Perhaps, only when the whole world is being defined as *one global knowledge cluster and innovation network*, then, for the moment, we cannot aggregate and escalate further to a mega-cluster or mega-network.

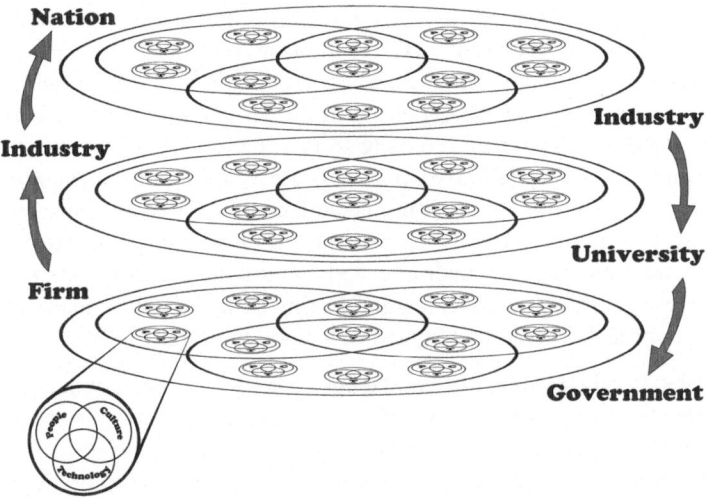

Fig. 12 The twenty-first century fractal innovation ecosystem. Source: Derived from Authors' unpublished notes and lectures at GWU, Authors' own conceptualization, adapted from Carayannis and Campbell (2009, p. 223)

5. *The adaptive integration and co-evolution of different knowledge and innovation modes, the "Quadruple Helix," and "Quintuple Helix"*: "Mode 3" allows and emphasizes the co-existence and co-evolution of different knowledge and innovation paradigms. In fact, a key hypothesis is: *The competitiveness and superiority of a knowledge system or the degree of advanced development of a knowledge system are highly determined by their adaptive capacity to combine and integrate different knowledge and innovation modes via co-evolution, co-specialization, and co-opetition knowledge stock and flow dynamics* (e.g., Mode 1, Mode 2, Triple Helix, and linear and nonlinear innovation). The specific context (circumstances, demands, configurations, and cases) determines which knowledge and innovation mode (*multimodal*), at which level (*multilevel*), involving what parties or agents (*multilateral*) and with what knowledge nodes or knowledge clusters (*multinodal*) will be appropriate. What results is an emerging fractal knowledge and innovation ecosystem ("Mode 3 Innovation Ecosystem"), well-configured for the knowledge economy and society challenges and opportunities of the twenty-first century by being endowed with mutually complementary and reinforcing as well as dynamically co-evolving, co-specializing, and co-opeting diverse and heterogeneous configurations of knowledge creation, diffusion, and use. The intrinsic litmus test of the capacity of such an ecosystem to survive and prosper in the context of continually gloCalizing and intensifying competition represents the ultimate competitiveness benchmark with regards to the robustness and quality of the ecosystem's knowledge and innovation architecture and

topology as it manifests itself in the form of a knowledge value-adding chain. The concept of the "Quadruple Helix" innovation systems broadens our understanding, because it adds the "media-based and culture-based public" and "civil society" to the picture. The "Quintuple Helix" is even broader, by contextualizing the Quadruple Helix by referring to the "natural environments of society" (Carayannis and Campbell 2010, p. 62). The *FREIE* represents another conceptual view of bringing those different and complex perspectives dynamically together, what is necessary, when we want to understand, manage, and govern Mode 3 as well as the Quadruple and Quintuple Helices. Open Innovation Diplomacy qualifies as a novel and interesting strategy, policy-making, and governance approach in context of Quadruple and Quintuple Helix.

The societal embeddedness of knowledge represents a theme that already Mode 2 and Triple Helix explicitly acknowledge. As a last thought for this article we want to underscore *the potentially beneficial cross-references between democracy and knowledge* for a better understanding of knowledge. In an attempt to define democracy, democracy could be shortcut as an interplay of two principles (Campbell 2005): (1) *Democracy can be seen as a method or procedure*, based on the application of the rule of the majority.[46] This acknowledges the "relativity of truth" and "pluralism" in a society, implying that decisions are carried out, not because they are "true" (or truer), but because they are backed and legitimized by a majority. Since, over time, these majority preferences normally shift, this creates political swings, driving the government/opposition cycles, which crucially add to the viability of a democratic system. (2) *Democracy can also be understood as a substance ("substantially")*, where substance, for example, is being understood as an evolutionary manifestation of fundamental rights (O'Donnell 2004, pp. 26–27, 47, 54–55). Obviously, the method/procedure and the substance approach overlap. Without fundamental rights, the majority rule could neutralize or even abolish itself. On the other hand, the practical "real political" implementation of rights also demands a political method, an institutionally setup procedure.

There are several international initiatives, interested in systematically measuring democracies in a global perspective and in empirical terms. These measurements allow drawing comparisons between theory of democracy and the actual behavior and performance of democracies. Freedom House, as an example, focuses on freedom as a key dimension of democracy, distinguishing between free, partly free and not free countries.[47] For Guillermo O'Donnell (2004), the interplay of human rights and human development defines and creates the quality of democracy. The Democracy Ranking, another democracy measurement initiative, is being theoretically influenced by O'Donnell and applies the following conceptual formula for defining the

---

[46] For example, Joseph A. Schumpeter (1942, Chaps. XX–III) emphasized this method-based criterion for democracy.

[47] For more information on Freedom House, see on the internet: http://www.freedomhouse.org/template.cfm?page=1.

quality of democracy: "quality of democracy = (freedom + other characteristics of the political system) + (performance of the non-political dimensions)" (Campbell 2008, p. 41).[48] Furthermore, the Democracy Ranking distinguishes between the following five dimensions: politics, gender, economy, knowledge, health, and the environment. To the dimension of politics a weight of 50% is assigned (for the overall ranking scores), all the other dimensions follow with a weight of 10% (Campbell 2008, pp. 33–34). With this focus on performance across a variety of dimensions, the Democracy Ranking wants to be *left/right neutral*, as far as possible, not favoring one-sidedly "freedom" or "equality." Often, freedom is being associated more closely to conservative (right) and equality to left ideologies (Campbell 2008, pp. 31–32; see also Campbell and Barth 2009). The Democracy Ranking asserts conceptually a link between quality of democracy and "sustainable development" (at least in a mid-term or long-term perspective). Furthermore, with the specific selection of dimensions for their model of democracy and the quality of democracy, the Democracy Ranking emphasizes knowledge (and innovation) and the environment (the natural environments of society). This makes the Democracy Ranking clearly Quadruple Helix-friendly and also Quintuple Helix-friendly, supporting comparative analysis of democracy, knowledge, and innovation. Some key findings of the *Democracy Ranking 2010* are (Campbell 2010, p. 2): "The Nordic countries (Norway, Sweden, Finland, Denmark) and Switzerland are the top five countries, also New Zeeland, the Netherlands, Ireland, Germany, and the UK have very high scores. This continuing global top position of the Nordic countries is impressive, also because this top position is being reproduced quite stable across the different (sub-)dimensions. Thus it can be said that the Nordic countries define—in a positive view—a global benchmark for quality of democracy that is empirically already available. From the top 10 countries seven belong to the EU. In total, the prominent representation of European democracies at the top positions is remarkable. This underscores that the European integration process should be understood, in the global context, even more clearly as a 'democracy project.'" Sustainable development, progress, and performance across different dimensions provide one explanation for the success and the high quality of democracy in the Nordic countries. These are some of the lessons to be learned in context of global analysis (see also Barth 2010).

Linking democracy even more directly to knowledge and innovation, we want to highlight the following aspects (see Fig. 13 for a suggested first-attempt graphical visualization; see also Godoe 2007, p. 358; and Carayannis and Ziemnowicz 2007):

1. *Knowledge-based and innovation-based democracy*: The future of democracy depends on evolving, enhancing, and ideally perfecting the concepts of a knowledge-based and innovation-based democratic polity as the manifestation and operationalization of what one might consider the, paraphrased, "twenty-first century platonic ideal state": "It has been basic United States policy that

---

[48] On the web, the Democracy Ranking can be visited under: http://www.democracyranking.org/en/.

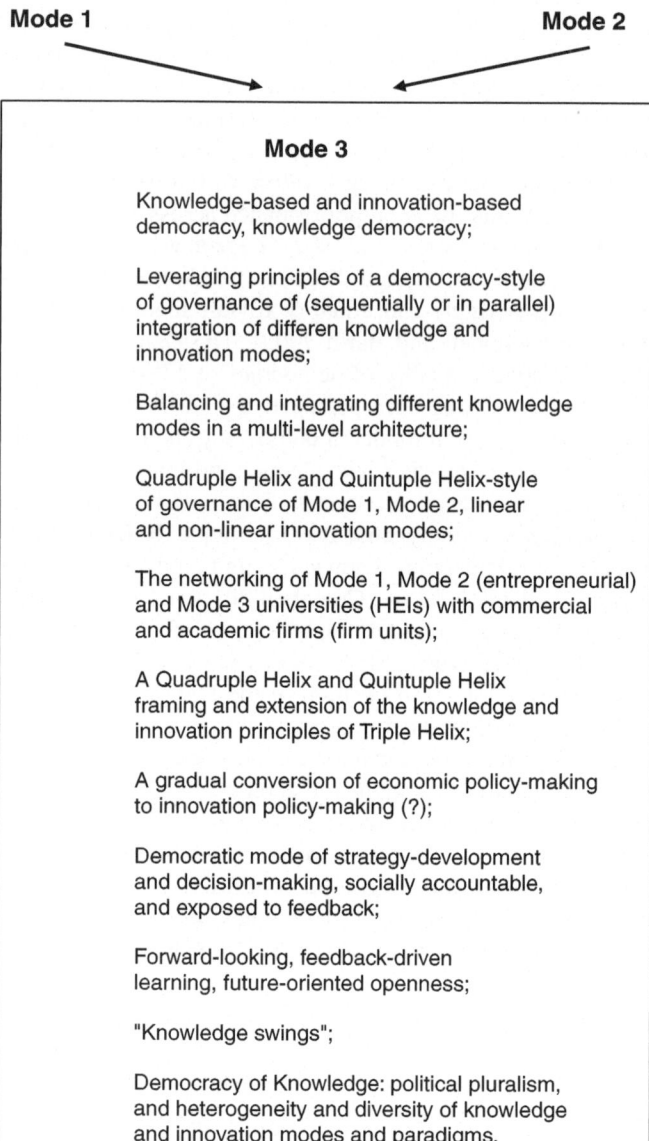

Mode 1   **Mode 2**

Mode 3

Knowledge-based and innovation-based democracy, knowledge democracy;

Leveraging principles of a democracy-style of governance of (sequentially or in parallel) integration of differen knowledge and innovation modes;

Balancing and integrating different knowledge modes in a multi-level architecture;

Quadruple Helix and Quintuple Helix-style of governance of Mode 1, Mode 2, linear and non-linear innovation modes;

The networking of Mode 1, Mode 2 (entrepreneurial) and Mode 3 universities (HEIs) with commercial and academic firms (firm units);

A Quadruple Helix and Quintuple Helix framing and extension of the knowledge and innovation principles of Triple Helix;

A gradual conversion of economic policy-making to innovation policy-making (?);

Democratic mode of strategy-development and decision-making, socially accountable, and exposed to feedback;

Forward-looking, feedback-driven learning, future-oriented openness;

"Knowledge swings";

Democracy of Knowledge: political pluralism, and heterogeneity and diversity of knowledge and innovation modes and paradigms.

Fig. 13 Knowledge, innovation and democracy in a Democracy of Knowledge: glocal governance styles of the Glocal Knowledge Economy and Society?. Source: Authors' own conceptualization based on Godoe (2007, p. 358) and on Carayannis and Campbell (2009, p. 226)

Government should foster the opening of new frontiers. It opened the seas to clipper ships and furnished land for pioneers. Although these frontiers have more or less disappeared, the frontier of science remains. It is in keeping with the American tradition—one which has made the United States great—that new frontiers shall be made accessible for development by all American citizens" (Bush 1945, p. 10). Knowledge, innovation, and democracy interrelate. Advances in democracy and advances in knowledge and innovation express mutual dependencies.[49] The "quality of democracy" depends on a knowledge base. We see how the Glocal Knowledge Economy and Society and the quality of democracy intertwine. Concepts, such as "democratizing innovation" (Von Hippel 2005), underscore such aspects. Also the media-based and culture-based public of the "Quadruple Helix" emphasizes the overlapping tendencies of democracy and knowledge.[50]

2. *Pluralism of knowledge modes*: Democracy's strength lies exactly in its capacity for allowing and balancing different parties, politicians, ideologies, values, and policies, and this ability was discussed by Lindblom (1959) as *disjointed incrementalism*[51]: "… as the partisan mutual adjustment process: Just as entrepreneurs and consumers can conduct their buying and selling without anyone attempting to calculate the overall level of prices or outputs for the economy as a whole, Lindblom argued, so in politics. Under many conditions, in fact, adjustments among competing partisans will yield more sensible policies than are likely to be achieved by centralized decision makers relying on analysis (Lindblom 1959, 1965). This is partly because interaction economizes on precisely the factors on which humans are short, such as time and understanding, while analysis requires their profligate consumption. To put this differently, the lynchpin of Lindblom's thinking was that analysis could be—and should be—no more than an adjunct to interaction in political life" (http://www.rpi.edu/~woodhe/docs/redner.724.htm). Similarly, democracy enables the integrating, coexistence, and co-evolution of different knowledge and innovation modes. We can speak of a pluralism of knowledge modes, and can regard this as a competitiveness feature of the whole system. Different knowledge modes can be linked to different knowledge decisions and knowledge policies, reflecting the communication skills of specific knowledge producers and knowledge users to convince other audiences of decision makers.

3. *"Knowledge swings"*: Through political cycles or *political swings* (Campbell 1992) a democracy ties together different features: (a) decides, who currently governs; (b) gives the opposition a chance, to come to power in the future; (c) and acknowledges pluralism. Democracy represents a system which always creates and is being driven by an important momentum of dynamics. For example,

---

[49] For attempts, trying to analyze the quality of a democracy, see for example Campbell and Schaller (2002).

[50] On "democratic innovation," see, furthermore, Saward (2006).

[51] The *disjointed incrementalism approach* to decision making (also known as *partisan mutual adjustment*) was developed by Lindblom (1959, 1965) and Linblom and Cohen (1979) and found several fields of application and use: "The Incrementalist approach was one response to the challenge of the 1960s. This is the theory of Charles Lindblom, which he described as 'partisan mutual adjustment' or disjointed incrementalism. Developed as an alternative to RCP, this theory claims that public policy is actually accomplished through decentralized bargaining in a free market and a democratic political economy" (http://www3.sympatico.ca/david.macleod/PTHRY.HTM).

the statistical probability for governing parties to lose an upcoming election is higher than to win an election (Müller and Strøm 2000, p. 589). Similarly, one could paraphrase the momentum of political swings by referring to "knowledge swings": in certain periods and concrete contexts, a specific set of knowledge modes expresses a *"dominant design"*[52] position; however, also the pool of non-hegemonic knowledge modes is necessary, for allowing alternative approaches in the long run, adding crucially to the variability of the whole system. "Knowledge swings" can have at least two ramifications: (a) What are dominant and nondominant knowledge modes in a specific context? (b) There is a pluralism of knowledge modes, which exist in parallel, and thus also codevelop and coevolve. Diversity is necessary to draw a cyclically patterned dominance of knowledge modes.

4. *Forward-looking, feedback-driven learning*: Democracy should be regarded as a future-oriented governance system, fostering and relying upon social, economic, and technological learning. The "Mode 3 Innovation Ecosystem" is at its foundation an open, adaptive, learning-driven knowledge, and innovation ecosystem reflecting the philosophy of *Strategic or Active Incrementalism* (Carayannis 1993, 1994, 1999, 2000, 2001) and the strategic management of technological learning (Carayannis 1999; see, furthermore, De Geus 1988). In addition, one can postulate that the government/opposition cycle in politics represents a feedback-driven learning and mutual adaptation process. In this context, a democratic system can be perceived of as a pendulum with a shifting pivot point reflecting the evolving, adapting dominant worldviews of the polity as they are being shaped by the mutually interacting and influencing citizens and the dominant designs of the underlying cultures and technological paradigms (Carayannis 2001, pp. 26–27).

In conclusion, we have attempted to provide an emerging conceptual framework to serve as the "intellectual sandbox" and "creative whiteboard space" of the mind's eyes of "knowledge weavers" (*Wissensweber*)[53] across disciplines and sectors as they strive to tackle the twenty-first century challenges and opportunities for socio–economic prosperity and cultural renaissance based on knowledge and innovation: "As a result of the glocalized nature and dynamics of state-of-the-art, specialized knowledge … one needs to cope with and leverage two mutually-reinforcing and complementary trends: (a) the symbiosis and co-evolution of top-down national and multi-national science, technology and innovation public policies … and bottom-up technology development and knowledge acquisition private initiatives; and (b) the leveling of the competitive field across regions of the world via technology diffusion

---

[52] "Studies have shown that the early period of a new area of technology is often characterized by technological ferment but that the pace of change slows after the emergence of a dominant design" (http://www.findarticles.com/p/articles/mi_m4035/is_1_45/ai_63018122/print).

[53] The term constitutes the brainchild or *conceptual branding* of the authors as part of this journey of discovery and ideation.

and adoption accompanied and complemented by the formation and exacerbation of multi-dimensional, multi-lateral, multi-modal and multi-nodal divides (cultural, technological, socio–economic, …) …In closing, being able to practice these two functions—being able to be a superior manager and policy-maker in the twenty-first century—relies on a team's, firm's, or society's capacity to be superior learners … in terms of both learning new facts as well as adopting new rules for learning-how-to-learn and establishing superior strategies for learning to learn-how-to-learn. Those superior learners will, by necessity, be both courageous and humble as these virtues lie at the heart of successful learning" (Carayannis and Alexander 2006). Already the early Lundvall (1992, pp. 1, 9) underscored the importance of learning for every national innovation system.

Mode 3 (Mode 3 knowledge production), in combination with the widened perspective of the Quadruple Helix and Quintuple Helix (Quadruple and Quintuple Helices innovation systems), emphasizes an Innovation Ecosystem (social and natural systems and environments) that encourages the co-evolution of different knowledge and innovation modes as well as balances nonlinear innovation modes in the context of multilevel innovation systems. Hybrid innovation networks and knowledge clusters tie together universities, commercial firms, and academic firms. Mode 3 may indicate an evolutionary and learning-based escape route for Schumpeter's "creative destruction" (Carayannis and Ziemnowicz 2007). The "knowledge state" (Campbell 2006b) has the potential to network "high-quality" democracy with the gloCal knowledge economy and society. There appears to be, at least potentially, a co-evolution and congruence between advanced knowledge, innovation, economy, society, and democracy. The *Democracy of Knowledge*, as a concept and metaphor, highlights and underscores parallel processes between political pluralism in advanced democracy, and knowledge and innovation heterogeneity and diversity in advanced economy and society. Here, we may observe a hybrid overlapping between the *knowledge economy, knowledge society, and knowledge democracy* (see again Figs. 7 and 13). High-quality democracies encourage *sustainable development* across a broad spectrum of dimensions, where, for certain, knowledge and innovation are of a key importance. High-quality democracies are "broader" than earlier concepts of a liberal democracy that were restricted to electoral democracy. What are necessary, but also possible *innovations in democracy*, to support the formation of *high-quality knowledge democracies*?[54] There is even more of a tendency that democracy as well as processes of advancing knowledge and innovation will become continuously broader, conceptually and in empirical terms (Carayannis and Campbell 2010, pp. 54–58, 60–61). We encourage to seeing the creative spectrum of the manifold links and cross-links between *innovation, entrepreneurship, and democracy*.

---

[54] View again "Democratic Innovation," edited by Michael Saward (2006). There is always a need to evaluate different policy fields of politics in reference to the values of democracy and the quality of democracy, for example asylum policy (Rosenberger 2010).

References

Anbari, Frank T., Stuart A. Umpleby (2006). Productive Research Teams and Knowledge Generation, 26–38, in: Elias G. Carayannis, David F. J. Campbell (eds.): Knowledge Creation, Diffusion, and Use in Innovation Networks and Knowledge Clusters. A Comparative Systems Approach across the United States, Europe and Asia. Westport, Connecticut: Praeger.

Arnold, Markus (2009). Interdisziplinarität: Theorie und Praxis eines Forschungskonzepts, 65–97, in: Markus Arnold (ed.): iff. Interdisziplinäre Wissenschaft im Wandel. Vienna: LIT.

Barth, Thorsten D. (2010). Konzeption, Messung und Rating der Demokratiequalität. Brasilien, Südarfika, Australien und die Russische Föderation, 1997–2006. Saarbrücken: VDM Verlag Dr. Müller.

Ben-Ari, Guy (2006). Innovation Policy in the Knowledge-Based Economy: The Israeli Case, 253–282, in: Elias G. Carayannis, David F. J. Campbell (eds.): Knowledge Creation, Diffusion, and Use in Innovation Networks and Knowledge Clusters. A Comparative Systems Approach across the United States, Europe and Asia. Westport, Connecticut: Praeger.

Biegelbauer, Peter (ed.) (2010). Steuerung von Wissenschaft? Die Governance des österreichischen Innovationssystems. Innsbruck: Studienverlag.

Blimlinger, Eva, Marcus Bruckmann, David F. J. Campbell, Bernhard Kernegger, Verena Krieger, Susanne Mann, Ruth Mateus-Berr, Barbara Putz-Plecko, Karin Raith, Emma Rendl-Denk, Veroniks Schnell, Maria Wiala (2010). Teaching, Quality, Evaluation. An Applied Concept. Vienna: University of Applied Arts Vienna (http://www.uni-ak.ac.at/uqe/download/TeachingEvaluation_AppliedConcept.pdf).

Brandenburger, Adam M., Barry J. Nalebuff (1997). Co-Opetition. New York: Doubleday.

Bush, Vannevar (1945). Science: The Endless Frontier. Washington, D.C.: United States Government Printing Office [http://www.nsf.gov/od/lpa/nsf50/vbush1945.htm#transmittal].

Braun, C. F. von (1997). The Innovation War. Upper Saddle River, NJ: Prentice Hall.

Caduff, Corina, Fiona Siegenthaler, Tan Wälchli (2010). Art and Artistic Research. Zurich Yearbook oft he Arts. Zurich: Zurich Universiy of the Arts.

Campbell, David F. J. (1992). Die Dynamik der politischen Links-Rechts-Schwingungen in Österreich: Die Ergebnisse einer Expertenbefragung. Österreichische Zeitschrift für Politikwissenschaft 21 (2), 165–179.

Campbell, David F. J. (1994). European Nation-State under Pressure: National Fragmentation or the Evolution of Suprastate Structures? Cybernetics and Systems: An International Journal 25 (6), 879–909.

Campbell, David F. J. (1995). Forschung und Forschungspolitik in Österreich. Ein strategisches Aktionsprogramm für die Sozialwissenschaften. SWS-Rundschau 35 (4), 395–404.

Campbell, David F. J. (1999). Evaluation universitärer Forschung. Entwicklungstrends und neue Strategiemuster für wissenschaftsbasierte Gesellschaften. SWS-Rundschau 39 (4), 363–383.

Campbell, David F. J. (2000). Forschungspolitische Trends in wissenschaftsbasierten Gesellschaften. Strategiemuster für entwickelte Wirtschaftssysteme. Wirtschaftspolitische Blätter 47 (2), 130–143.

Campbell, David F. J. (2001). Politische Steuerung über öffentliche Förderung universitärer Forschung? Systemtheoretische Überlegungen zu Forschungs- und Technologiepolitik. Österreichische Zeitschrift für Politikwissenschaft 30 (4), 425–438.

Campbell, David F. J. (2003). The Evaluation of University Research in the United Kingdom and the Netherlands, Germany and Austria, 98–131, in: Philip Shapira, Stefan Kuhlmann (eds.): Learning from Science and Technology Policy Evaluation: Experiences from the United States and Europe. Camberley: Edward Elgar.

Campbell, David F. J. (2005). Demokratie, Demokratiequalität und Grundrechte: Ein Vergleich der Fiedler- und EU-Verfassung. Vienna: Unpublished Manuscript.

Campbell, David F. J. (2006a). The University/Business Research Networks in Science and Technology: Knowledge Production Trends in the United States, European Union and Japan, 67–100, in: Elias G. Carayannis, David F. J. Campbell (eds.): Knowledge Creation, Diffusion, and Use in Innovation Networks and Knowledge Clusters. A Comparative Systems Approach across the United States, Europe and Asia. Westport, Connecticut: Praeger.

Campbell, David F. J. (2006b). Nationale Forschungssysteme im Vergleich. Strukturen, Herausforderungen und Entwicklungsoptionen. *Österreichische Zeitschrift für Politikwissenschaft* 35 (1), 25–44. [http://www.oezp.at/oezp/online/online.htm].

Campbell, David F. J. (2008). The Basic Concept for the Democracy Ranking of the Quality of Democracy. Vienna: Democracy Ranking (http://www.democracyranking.org/downloads/basic_concept_democracy_ranking_2008_A4.pdf).

Campbell, David F. J. (2009). "Externe Umwelten". Außensichten auf das iff, 99–134, in: Markus Arnold (ed.): iff. Interdisziplinäre Wissenschaft im Wandel. Vienna: LIT.

Campbell, David F. J. (2010). Key Findings (Summary Abstract) of the Democracy Ranking 2010 and the Democracy Improvement Ranking 2010. Vienna: Democracy Ranking (http://www.democracyranking.org/downloads/Key%20findings%20of%20the%20Democracy%20Ranking%202010_A4.pdf).

Campbell, David F. J. (2011). Wissenschaftliche „Parallelkarrieren" als Chance. Wenn Wissenschaft immer öfter zur Halbtagsbeschäftigung wird, könnte eine Lösung im „Cross-Employment" liegen. Guest Commentary for DIE PRESSE (February 2, 2011) (http://diepresse.com/home/bildung/meinung/635781/Wissenschaftliche-Parallelkarrieren-als-Chance?direct=635777&_vl_backlink=/home/bildung/index.do&selChannel=500).

Campbell, David F. J., Christian Schaller (eds.) (2002). Demokratiequalität in Österreich. Zustand und Entwicklungsperspektiven. Opladen: Leske + Budrich. [http://www.oegpw.at/sek_agora/publikationen.htm].

Campbell, David F. J., Wolfgang H. Güttel (2005). Knowledge Production of Firms: Research Networks and the "Scientification" of Business R&D. *International Journal of Technology Management* 31 (1/2), 152–175.

Campbell, David F. J., Thorsten D. Barth (2009). Wie können Demokratie und Demokratiequalität gemessen werden? Modelle, Demokratie-Indices und Länderbeispiele im globalen Vergleich. (How Can Democracy and the Quality of Democracy be Measured? Models, Democracy Indices and Country-Based Case Studies in Global Comparison.) *SWS-Rundschau* 49 (2), 209–233 (http://www.ssoar.info/ssoar/View/?resid=12471).

Campbell, George S., David F. J. Campbell (2011). The Semi-Aquatic Theory: Semi-Aquatic Evolutionary Phase and Environment, Language Development of Modern Humans. With a Short Epilog on Conceptualized Evolution, Social Ecology and the Quintuple Helix. *International Journal of Social Ecology and Sustainable Development* 2 (1), 15–30 (http://www.igi-global.com/bookstore/article.aspx?titleid=51634).

Carayannis, Elias G. (1993). Incrementalisme Strategique. *Le Progrès Technique* (no. 2), Paris: France.

Carayannis, Elias G. (1994). Gestion Strategique de l'Apprentissage Technologique. *Le Progrès Technique* (no. 2), Paris: France.

Carayannis, Elias G. (1997). Data warehouse, electronic commerce, and technological learning: Successes and failures from government and private industry and lessons learned for 21st century electronic government. *Online Journal of Internet Banking and Commerce* (March).

Carayannis, Elias G. (1998). Higher order technological learning as determinant of market success in the multimedia arena: A success story, a failure, and a question to mark: Agfa/Bayer AG, enable software, and sun microsystems. *International Journal of Technovation* 18 (10), 639–653.

Carayannis, Elias G. (2008). Knowledge-driven creative destruction, or leveraging knowledge for competitive advantage: strategic knowledge arbitrage and serendipity as real options drivers triggered by co-opetition, co-evolution and co-specialization. Industry and Higher Education 22 (6), 343–353.

Carayannis, Elias G. (1999). Knowledge Transfer through Technological Hyperlearning in Five Industries. *International Journal of Technovation* 19 (3, March), 141–161.

Carayannis, Elias G. (2000). Investigation and Validation of Technological Learning versus Market Performance. *International Journal of Technovation* 20 (7, July), 389–400.

Carayannis, Elias G. (2001). The Strategic Management of Technological Learning. Learning to Learn and Learning to Learn-How-To-Learn as Drivers of Strategic Choice and Firm Performance in Global, Technology-Driven Markets. Boca Raton, Florida: CRC Press.

Carayannis, Elias G. (2004). Measuring Intangibles: Managing Intangibles for Tangible Outcomes in Research and Innovation. *International Journal of Nuclear Knowledge Management,* v. 1, no. 1, January.

Carayannis, Elias G., Jeffrey Alexander (1999a). Winning by Co-opeting in Strategic Government-University-Industry (GUI) Partnerships: The Power of Complex, Dynamic Knowledge Networks. *Journal of Technology Transfer* 24 (2/3, August), 197–210.

Carayannis, Elias G., Jeffrey Alexander (1999b). Technology-Driven Strategic Alliances: Tools for Learning and Knowledge Exchange in a Positive-Sum World, 1–32 until 1–41, in: Richard C. Dorf (ed.): The Technology Management Handbook. Boca Raton, Florida: CRC Press.

Carayannis, Elias G., Edgar Gonzalez (2003). Creativity and Innovation = Competitiveness? When, How, and Why, Vol. 1, Chap. 8, pp. 587–606, in: Larisa V. Shavinina (ed.): The International Handbook on Innovation. Amsterdam: Pergamon.

Carayannis, Elias G., Edgar Gonzalez, John Wetter (2003). The Nature and Dynamics of Discontinuous and Disruptive Innovations From a Learning and Knowledge Management Perspective, Vol. 1, Chap. 4, pp. 115–138, in: Larisa V. Shavinina (ed.): The International Handbook on Innovation. Amsterdam: Pergamon.

Carayannis, Elias G., Jeffrey Alexander (2004). Strategy, Structure and Performance Issues of Pre-competitive R&D Consortia: Insights and Lessons Learned. *IEEE Transactions of Engineering Management* 52 (2).

Carayannis, Elias G., Patrice Laget (2004). Transatlantic Innovation Infrastructure Networks: Public-Private, EU-US R&D Partnerships. *R&D Management* 34 (1), 17–31.

Carayannis, Elias G., Maximilian von Zedtwitz (2005). Architecting GloCal (Global – Local), Real-Virtual Incubator Networks (G-RVINs) as Catalysts and Accelerators of Entrepreneurship in Transitioning and Developing Economies. *Technovation* 25, 95–110.

Carayannis, Elias G., Jeffrey M. Alexander (2006). Global and Local Knowledge. Glocal Transatlantic Public-Private Partnerships for Research and Technological Development. Houndmills: Palgrave MacMillan.

Carayannis, Elias G., David F. J. Campbell (2006a). „Mode 3": Meaning and Implications from a Knowledge Systems Perspective, 1–25, in: Elias G. Carayannis, David F. J. Campbell (eds.): Knowledge Creation, Diffusion, and Use in Innovation Networks and Knowledge Clusters. A Comparative Systems Approach across the United States, Europe and Asia. Westport, Connecticut: Praeger.

Carayannis, Elias G., David F. J. Campbell (2006b). Conclusion: Key Insights and Lessons Learned for Policy and Practice, 331–341, in: Elias G. Carayannis, David F. J. Campbell (eds.): Knowledge Creation, Diffusion, and Use in Innovation Networks and Knowledge Clusters. A Comparative Systems Approach across the United States, Europe and Asia. Westport, Connecticut: Praeger.

Carayannis, Elias G., David F. J. Campbell (2006c). Introduction and Chapter Summaries, ix-xxvi, in: Elias G. Carayannis, David F. J. Campbell (eds.): Knowledge Creation, Diffusion, and Use in Innovation Networks and Knowledge Clusters. A Comparative Systems Approach across the United States, Europe and Asia. Westport, Connecticut: Praeger.

Carayannis, Elias G., Caroline Sipp (2006). E-Development toward the Knowledge Economy: Leveraging Technology, Innovation and Entrepreneurship for "Smart Development". Houndmills: Palgrave MacMillan.

Carayannis, Elias G., Christopher Ziemnowicz (eds.) (2007). Rediscovering Schumpeter. Creative Destruction Evolving into "Mode 3". Houndmills: Palgrave MacMillan.

Carayannis, Elias G., John E. Spillan, Christopher Ziemnowicz (2007). Introduction: Why Joseph Schumpeter's Creative Destruction? Everything has Changed, 1–5, in: Elias G. Carayannis, Christopher Ziemnowicz (eds.): Rediscovering Schumpeter. Creative Destruction Evolving into "Mode 3". Houndmills: Palgrave MacMillan.

Carayannis, Elias G., David F. J. Campbell (2009). "Mode 3" and "Quadruple Helix": Toward a 21st Century Fractal Innovation Ecosystem. *International Journal of Technology Management*

46 (3/4), 201–234 (http://www.inderscience.com/browse/index.php?journalID=27&year=200 9&vol=46&issue=3/4).

Carayannis, Elias G, David F. J. Campbell (2010). Triple Helix, Quadruple Helix and Quintuple Helix and How Do Knowledge, Innovation and the Environment Relate To Each Other? A Proposed Framework for a Trans-disciplinary Analysis of Sustainable Development and Social Ecology. *International Journal of Social Ecology and Sustainable Development* 1 (1), 41–69 (http://www.igi-global.com/bookstore/article.aspx?titleid=41959).

Carayannis, Elias G., David F. J. Campbell (2011). Open Innovation Diplomacy and a 21st Century Fractal Research, Education and Innovation (FREIE) Ecosystem: Building on the Quadruple and Quintuple Helix Innovation Concepts and the "Mode 3" Knowledge Production System. *Journal of the Knowledge Economy* 2 (3), 327–372 (http://www.springerlink.com/content/d1lr223321305579/).

Carayannis, Elias G (2012). The Knowledge of Culture and the Culture of Knowledge. Houndmills: Palgrave MacMillan (forthcoming).

Cardullo, Mario W. (1999). Technology Life Cycles, 3–44 until 3–49, in: Richard C. Dorf (ed.): The Technology Management Handbook. Boca Raton, Florida: CRC Press.

Colapinto, Cinzia, Colin Porlezza (2012). Innovation in Creative Industries: from the Quadruple Helix Model to the Systems Theory. *Journal of the Knowledge Economy* 3 (1) (forthcoming) (http://www.springerlink.com/content/rx725r81u91199g5/).

Cooper, J. R. (1998). A Multidimensional Approach to the Adoption of Innovation. *Management Decision* 36 (8), 493–502.

Cesaroni, Fabrizio, Alfonso Gambardella, Walter Garcia-Fontes, Myriam Mariani (2004). The Chemical Sectoral System: Firms, Markets, Institutions and the Processes of Knowledge Creation and Diffusion, 121–154, in: Malerba, Franco (ed.): Sectoral Systems of Innovation. Concepts, Issues and Analyses of Six Major Sectors in Europe. Cambridge: Cambridge University Press.

Davenport, Thomas H., John C. Beck (2001). The Attention Economy. Understanding the New Currency of Business. Boston, Massachusetts: Harvard Business School Press.

Danilda, Inger, Malin Lindberg, Britt-Marie Torstensson (2009). Women Resource Centres. A Quattro Helix Innovation System on the European Agenda. Paper (http://www.hss09.se/own_documents/Papers/3-11%20-%20Danilda%20Lindberg%20&%20Torstensson%20-%20paper.pdf).

De Geus, A. (1988). Planning as Learning, Harvard Business Review, 66:2, 70, Winter.

Drejer, Anders (2002). Strategic Management and Core Competencies.New York, NY: Quorum Books.

Dubina, Igor N., Elia G. Carayannis, David F. J. Campbell (2012). Creativity Economy and a Crisis of the Economy? Coevolution of Knowledge, Innovation, and Creativity, and of the Knowledge Economy and Knowledge Society. *Journal of the Knowledge Economy* 3 (1) (forthcoming) (http://www.springerlink.com/content/t5j8l12136h526h5/).

Esping-Andersen, Gøsta (1990). The Three Worlds of Welfare Capitalism. Princeton, New Jersey: Princeton University Press.

Etzkowitz, Henry, Loet Leydesdorff (2000). The dynamics of innovation: from National Systems and "Mode 2" to a Triple Helix of university-industry-government relations. *Research Policy* 29, 109–123.

Etzkowitz, Henry (2003). Research groups as "quasi-firms": The invention of the Entrepreneurial University. *Research Policy* 32, 109–121.

Ferlie, Ewan, Christine Musselin, Gianluca Andresani (2008). The Steering of Higher Education Systems: A Public Management Perspective. *Higher Education* 56 (3), 325–348 (http://www.springerlink.com/content/n22v788851377144/fulltext.pdf).

Ferlie, Ewan, Christine Musselin, Gianluca Andresani (2009). The Governance of Higher Education Systems: A Public Management Perspective, 1–20, in: Catherine Paradeise, Emanuela Reale, Ivar Bleiklie, Ewan Ferlie (eds.). University Governance. Western European Comparative Perspectives. Dordrecht: Springer.

Fischer-Kowalski, Marina, Helmut Haberl (eds.) (2007). Socioecological Transitions and Global Change. Trajectories of Social Metabolism and Land Use. Cheltenham: Edward Elgar.

Florida, Richard (2004). The Rise of the Creative Class: And How It's Transforming Work, Leisure, Community, and Everyday Life. Cambridge, MA: Basic Books.

Gerybadze, Alexander, Guido Reger (1999). Globalization of R&D: Recent Changes in the Management of Innovation in Transnational Corporations. *Research Policy* 28, 251–274.

Geuna, Aldo, Ben R. Martin (2003). University Researech Evaluation and Funding: An International Comparison. *Minerva* 41, 277–304.

Gibbons, Michael, Camille Limoges, Helga Nowotny, Simon Schwartzman, Peter Scott, Martin Trow (1994). The New Production of Knowledge. The Dynamics of Science and Research in Contemporary Societies. London: Sage.

Gleick, James (1987). Chaos: Making a New Science. New York: Viking Press.

Godoe, Helge (2007). Doing Innovative Research: "Mode 3" and Methodological Challenges in Leveraging the Best of Three Worlds, 344–361, in: Elias G. Carayannis, Christopher Ziemnowicz (eds.): Rediscovering Schumpeter. Creative Destruction Evolving into "Mode 3". Houndmills: Palgrave MacMillan.

Gottweis, Herbert (1998). Governing Molecules. The Discursive Politics of Genetic Engineering in Europe and the United States. Cambridge, Massachusetts: MIT Press.

Hall, Peter A. (1993). Policy Paradigms, Social Learning, and the State. The Case of Economic Policymaking in Britain. *Comparative Politics* (April 1993), 257–296.

Hemlin, Sven, Carl Martin Allwood, Ben R. Martin (2004). Creative Knowledge Environments. The Influences on Creativity in Research and Innovation. Cheltenham: Edward Elgar.

Hindmarsh, Richard, Barbara Prainsack (eds.). Genetic Suspects. Global Governance of Forensic DNA Profiling and Databasing. Cambridge, UK: Cambridge University Press.

Hooghe, Liesbet, Gary Marks (2001). Multi-Level Governance and European Integration. Lanham: Rowman & Littlefield Publishers.

Jacob, Anna Katharina (2007). Qualitätsmanagement an Musikhochschulen in Zeiten sich wandelnder Studienstrukturen. Hildesheim: Olms.

Kaiser, Robert, Heiko Prange (2004). The Reconfiguration of National Innovation Systems – The Example of German Biotechnology. *Research Policy* 33, 395–408.

Killman, R. (1985). Gaining Control of the Corporate Culture. New York: McGraw-Hill.

Kline, S. J., N. Rosenberg (1986). An Overview of Innovation, in: R. Landau, N. Rosenburg (eds.): The Positive Sum Strategy. Washington, D.C.: National Academy Press.

Kritzinger, Sylvia, Barbara Prainsack, Helga Pülzl (2006). System oder Netzwerk? Veränderungen forschungspolitischer Strategien in Österreich. *Österreichische Zeitschrift für Politikwissenschaft* 35 (1), 75–92 (http://www.oezp.at/pdfs/2006-1-5-Kritzinge.pdf).

Krücken, Georg (2003a). Mission Impossible? Institutional Barriers to the Diffusion of the "Third Academic Mission" at German Universities. *International Journal of Technology Management* 25 (1/2), 18–33.

Krücken, Georg (2003b). Learning the "New, New Thing": On the Role of Path Dependency in University Structures. *Higher Education* 46 (3), 315–339.

Krücken Georg, Frank Meier, Andre Müller (2007). Information, Cooperation, and the Blurring of Boundaries – Technology Transfer in German and American Discourses. *Higher Education* 53 (6), 675–696.

Kuhlmann, Stefan (2001). Future Governance of Innovation Policy in Europe – Three Scenarios. *Research Policy* 30, 953–976.

Kuhn, Thomas S. (1962). The Structure of Scientific Revolutions. Chicago: The University of Chicago Press.

Leydesdorff, Loet (2012). The Triple Helix, Quadruple Helix, …, and an N-Tuple of Helices: Explanatory Models for Analyzing the Knowledge-Based Economy? *Journal of the Knowledge Economy* 3 (1) (forthcoming) (http://www.springerlink.com/content/x543613918677871/).

Lindberg, Malin, Inger Danilda, Britt-Marie Torstensson (2012). Women Resource Centres – A Creative Knowledge Environment of Quadruple Helix. *Journal of the Knowledge Economy* 3 (1) (forthcoming) (http://www.springerlink.com/content/t47q129240051g31/).

Lindblom, Charles E. (1959). The Science of Muddling Through. *Public Administration Review* 19, 79–88.

Lindblom, Charles E. (1965). The Intelligence of Democracy. New York: The Free Press.

Lindblom, Charles E., David K. Cohen (1979). Usable Knowledge: Social Science and Social Problem Solving. New Haven: Yale University Press.

Lundvall, Bengt-Åke (ed.) (1992). National Systems of Innovation. Towards a Theory of Innovation and Interactive Learning. London: Pinter Publishers.

Malerba, Franco (ed.) (2004a). Sectoral Systems of Innovation. Concepts, Issues and Analyses of Six Major Sectors in Europe. Cambridge: Cambridge University Press.

Malerba, Franco (2004b). Summing-Up and Conclusions, 465–507, in: Malerba, Franco (ed.): Sectoral Systems of Innovation. Concepts, Issues and Analyses of Six Major Sectors in Europe. Cambridge: Cambridge University Press.

McKelvey, Maureen, Luigi Orsenigo, Fabio Pammolli (2004). Pharmaceuticals Analyzed through the Lens of a Sectoral Innovation System, 73–120, in: Malerba, Franco (ed.): Sectoral Systems of Innovation. Concepts, Issues and Analyses of Six Major Sectors in Europe. Cambridge: Cambridge University Press.

McNiff, Shaun (1998). Art-Based Research. London: Jessica Kingsley.

McNiff, Shaun (2008). Art-Based Research, 29–40, in: J. Gary Knowles, Ardra L. Cole (eds.): Handbook of the Arts in Qualitative Research. Los Angeles: Sage.

Milbergs, Egils (2005). Innovation Ecosystems and Prosperity. Center for Accelerating Innovation [http://www.innovationecosystems.com].

Miyata, Yukio (2003). An Analysis of Research and Innovative Activities of Universities in the United States, 715–738, in: Larisa V. Shavinina (ed.): The International Handbook on Innovation. Amsterdam: Pergamon.

Müller, Wolfgang C., Kaare Strøm (2000). Conclusion: Coalition Governance in Western Europe, 559–592, in: Wolfgang C. Müller, Kaare Strøm (eds.): Coalition Governments in Western Europe.

National Science Board (2010). Science and Engineering Indicators 2010. Arlington, VA: National Science Foundation (http://www.nsf.gov/statistics/seind10/pdfstart.htm).

Nelson, Richard R. (ed.) (1993). National Innovation Systems. A Comparative Analysis. Oxford: Oxford University Press.

Nelson, Richard R., Sidney G., Winter (1982). An Evolutionary Theory of Economic Change. Cambridge, MA: Harvard University Press.

Nowotny, Helga, Peter Scott, Michael Gibbons (2001). Re-thinking science. Knowledge and the public in an age of uncertainty. Cambridge: Polity Press.

Nowotny, Helga, Peter Scott, Michael Gibbons (2003). Mode 2 Revisited: The New Production of Knowledge. Minerva 41, 179–194.

Nowotny, Helga, Peter Scott, Michael Gibbons (2006). Re-Thinking Science: Mode 2 in Societal Context, 39–51, in: Elias G. Carayannis, David F. J. Campbell (eds.): Knowledge Creation, Diffusion, and Use in Innovation Networks and Knowledge Clusters. A Comparative Systems Approach across the United States, Europe and Asia. Westport, Connecticut: Praeger.

O'Donnell, Guillermo (2004). Human Development, Human Rights, and Democracy, 9–92, in: Guillermo O'Donnell, Jorge Vargas Cullell, Osvaldo M. Iazzetta (eds.): The Quality of Democracy. Theory and Applications. Notre Dame, Indiana: University of Notre Dame Press.

OECD (1994). Frascati Manual. The Measurement of Scientific and Technological Acitivities. Proposed Standard Practice for Surveys of Research and Experimental Development. Paris: OECD.

OECD (1998). Science, Technology and Industry Outlook. Paris: OECD.

OECD (2002). Frascati Manual 2002. The Measurement of Scientific and Technological Activities. Proposed Standard Practice for Surveys on Research and Experimental Development. Paris: OECD (http://www.oecdbookshop.org/oecd/display.asp?CID=&LANG=EN&SF1=DI&ST1=5LMQCR2K61JJ).

OECD (2006). Research and Development Statistics. (On-Line Database). Paris: OECD.

Pechar, Hans, Lesley Andres (2011). Higher-Education Policies and Welfare Regimes: International Comparative Perspectives. Higher Education Policy 24 (March), 25–52 (http://www.palgrave-journals.com/hep/journal/v24/n1/full/hep201024a.html).

Pfeffer, Thomas (2006). Virtualization of Research Universities. Raising the Right Questions to Address Key Functions of the Institution, 307–330, in: Elias G. Carayannis, David F. J.

Campbell (eds.): Knowledge Creation, Diffusion, and Use in Innovation Networks and Knowledge Clusters. A Comparative Systems Approach across the United States, Europe and Asia. Westport, Connecticut: Praeger.

Plasser, Fritz (ed.) (2004). Politische Kommunikation in Österreich. Ein praxisnahes Handbuch. Vienna: WUV-Universitätsverlag.

Plasser, Fritz, Gunda Plasser (2002). Global Political Campaigning. A Worldwide Analysis of Campaign Professionals and Their Practices. Westport, Connecticut: Praeger.

Polanyi, Michael (1962). The Republic of Science: Its Political and Economic Theory. *Minerva* 1, 54–74 (http://sciencepolicy.colorado.edu/students/envs_5100/polanyi_1967.pdf and http://fiesta.bren.ucsb.edu/~gsd/595e/docs/41.%20Polanyi_Republic_of_Science.pdf).

Prainsack, Barbara, Howard Wolinsky (2010). Direct-to-Consumer Genome Testing: Opportunities for Pharmacogenomics Research? *Pharmacogenomics* 11 (5), 651–655.

Prem, Erich (2005). Zur Lage der österreichischen Forschung auf dem Gebiet integrierter Systeme. Situation und Strategie. *Elektrotechnik und Informationstechnik (ei)*.

Resetarits Andreas, Agnezia-Maria, Resetarits-Tincul (2012). Fuzzy Concepts – A New Approach in the Description of Boundaries as Creative Knowledge Environments in Educational Sciences. *Journal of the Knowledge Economy* 3 (1) (in press) (http://www.springerlink.com/content/j463335233513170/).

Ritterman, Janet, Gerald Bast, Jürgen Mittelstraß (eds.) (2011). Art and Research. Can Artists Be Researchers? Vienna: Springer.

Rogers, E. M., Shoemaker, F. F. (1971). *Communication of innovations: A cross-cultural approach.* New York: Free Press.

Rosenberger, Sieglinde (ed.) (2010). Asylpolitik in Österreich. Unterbringung im Fokus. Vienna: Facultas.

Rycroft, Robert W., Don E. Kash (1999). The Complexity Challenge. Technological Innovation for the 21st Century. London: Pinter.

Saward, Michael (ed.) (2006). Democratic Innovation: Deliberation, Representation and Association. London: Routledge.

Schumpeter, Joseph A. (1942). Capitalism, Socialism and Democracy. New York: Harper & Brothers.

Shapira, Philip, Stefan Kuhlmann (eds.) (2003). Learning from Science and Technology Policy Evaluation. Experiences from the United States and Europe. Cheltenham: Edward Elgar.

Shavinina, Larisa V. (2003). The International Handbook on Innovation. Amsterdam: Pergamon.

Steinmueller, W. Edward (2004). The European Software Sectoral System of Innovation, 193–242, in: Malerba, Franco (ed.): Sectoral Systems of Innovation. Concepts, Issues and Analyses of Six Major Sectors in Europe. Cambridge: Cambridge University Press.

Tassey, Gregory (2001). R&D Policy Models and Data Needs, 37–71, in: Maryann P. Feldman, Albert N. Link (eds.). Innovation Policy in the Knowledge-Based Economy. Boston: Kluwer Academic Publishers.

Teichler, Ulrich (2006). Was ist Qualität?, 168–184, in: Véronique Chalvet, Waldemar Dreger (eds.): Von der Qualitätssicherung der Lehre zur Qualitätsentwicklung als Prinzip der Hochschulsteuerung. Bonn: Hochschulkonferenz (Beiträge zur Hochschulpolitik).

Tornatzky, L.G., M. Fleischer. (1990). The Process of Technological Innovation. Lexington, MA: Lexington Books.

Umpleby, Stuart A. (1997). Cybernetics of Conceptual Systems. *Cybernetics and Systems: An International Journal* 28, 635–652.

Umpleby, Stuart A. (2002). Should knowledge of management be organized as theories or as methods?, 492–497, in: Robert Trappl (ed.): Cybernetics and systems 2002. Proceedings of the 16th European meeting on cybernetics and systems research. Volume 1. Vienna: Austrian Society for Cybernetic Studies.

Umpleby, Stuart A. (2005). What I Learned from Heinz von Foerster About the Construction of Science. *Kybernetes* 34 (1/2), 278–294.

Volkens, Andrea, Hans-Dieter Klingemann (2002). Parties, Ideologies, and Issues. Stability and Change in Fifteen European Party Systems 1945–1998, 143–167, in: Kurt Richard Luther,

Ferdinand Müller-Rommel (eds.): Political Parties in the New Europe. Political and Analytical Challenges. Oxford: Oxford University Press.

Von Hippel, Eric (1995). The Sources of Innovation. Oxford: Oxford University Press.

Von Hippel, Eric (2005). Democratizing Innovation. Cambridge, MA: MIT Press.

Von Zedtwitz, Max, Oliver Gassmann (2002). Market versus Technology Drive in R&D Internationalization: Four Different Patterns of Managing Research and Development. *Research Policy* 31 (4), 569–588.

Von Zedtwitz, Max, Philip Heimann (2006). Innovation in Clusters and the Liability of Foreignness of International R&D, 101–122, in: Elias G. Carayannis, David F. J. Campbell (eds.): Knowledge Creation, Diffusion, and Use in Innovation Networks and Knowledge Clusters. A Comparative Systems Approach across the United States, Europe and Asia. Westport, Connecticut: Praeger.

Wengel, Jürgen, Philip Shapira (2004). Machine Tools: The Remaking of a Traditional Sectoral Innovation System, 243–286, in: Malerba, Franco (ed.): Sectoral Systems of Innovation. Concepts, Issues and Analyses of Six Major Sectors in Europe. Cambridge: Cambridge University Press.

Winiwarter, Verena, Martin Knoll (2007) Umweltgeschichte. Köln: Böhlau.

Yau, Lynn Foon Chi (2012). The Arts in a Knowledge Economy: Creation of Other Knowledges. *Journal of the Knowledge Economy* 3 (1), (in press) (http://www.springerlink.com/content/n38t14j275250376/).

Zaltman, Gerald, Robert Ducan, Jonny Holbek (1973). Innovations and organizations. New York: Wiley.